CONTENTS

TESTIMONY

PREPARED STATEMENTS SUBMITTED BY MEMBERS OF CONGRESS

PREPARED STATEMENTS SUBMITTED FOR THE RECORD

SUBMISSIONS FOR THE RECORD

PREPAREDNESS AGAINST TERRORIST ATTACKS

(106–21)

HEARING

BEFORE THE

SUBCOMMITTEE ON

OVERSIGHT, INVESTIGATIONS AND EMERGENCY MANAGEMENT

OF THE

COMMITTEE ON TRANSPORTATION AND INFRASTRUCTURE

HOUSE OF REPRESENTATIVES

ONE HUNDRED SIXTH CONGRESS

FIRST SESSION

JUNE 9, 1999

Printed for the use of the
Committee on Transportation and Infrastructure

U.S. GOVERNMENT PRINTING OFFICE

60–220CC WASHINGTON : 1999

For sale by the U.S. Government Printing Office
Superintendent of Documents, Congressional Sales Office, Washington, DC 20402

ISBN 0-16-060644-6

COMMITTEE ON TRANSPORTATION AND INFRASTRUCTURE

BUD SHUSTER, Pennsylvania, *Chairman*

DON YOUNG, Alaska
THOMAS E. PETRI, Wisconsin
SHERWOOD L. BOEHLERT, New York
HERBERT H. BATEMAN, Virginia
HOWARD COBLE, North Carolina
JOHN J. DUNCAN, JR., Tennessee
THOMAS W. EWING, Illinois
WAYNE T. GILCHREST, Maryland
STEPHEN HORN, California
BOB FRANKS, New Jersey
JOHN L. MICA, Florida
JACK QUINN, New York
TILLIE K. FOWLER, Florida
VERNON J. EHLERS, Michigan
SPENCER BACHUS, Alabama
STEVEN C. LATOURETTE, Ohio
SUE W. KELLY, New York
RAY LAHOOD, Illinois
RICHARD H. BAKER, Louisiana
CHARLES F. BASS, New Hampshire
ROBERT W. NEY, Ohio
JACK METCALF, Washington
EDWARD A. PEASE, Indiana
ASA HUTCHINSON, Arkansas
MERRILL COOK, Utah
JOHN COOKSEY, Lousiana
JOHN R. THUNE, South Dakota
FRANK A. LOBIONDO, New Jersey
JERRY MORAN, Kansas
JOHN T. DOOLITTLE, California
LEE TERRY, Nebraska
DON SHERWOOD, Pennsylvania
GARY G. MILLER, California
JOHN E. SWEENEY, New York
JIM DEMINT, South Carolina
DOUG BEREUTER, Nebraska
STEVEN T. KUYKENDALL, California
MICHAEL K. SIMPSON, Idaho
JOHNNY ISAKSON, Georgia
DAVID VITTER, Louisiana

JAMES L. OBERSTAR, Minnesota
NICK J. RAHALL II, West Virginia
ROBERT A. BORSKI, Pennsylvania
WILLIAM O. LIPINSKI, Illinois
ROBERT E. WISE, JR., West Virginia
JAMES A. TRAFICANT, JR., Ohio
PETER A. DEFAZIO, Oregon
BOB CLEMENT, Tennessee
JERRY F. COSTELLO, Illinois
ELEANOR HOLMES NORTON, District of Columbia
JERROLD NADLER, New York
PAT DANNER, Missouri
ROBERT MENENDEZ, New Jersey
CORRINE BROWN, Florida
JAMES A. BARCIA, Michigan
BOB FILNER, California
EDDIE BERNICE JOHNSON, Texas
FRANK MASCARA, Pennsylvania
GENE TAYLOR, Mississippi
JUANITA MILLENDER-MCDONALD, California
ELIJAH E. CUMMINGS, Maryland
EARL BLUMENAUER, Oregon
MAX SANDLIN, Texas
ELLEN O. TAUSCHER, California
BILL PASCRELL, JR., New Jersey
LEONARD L. BOSWELL, Iowa
JAMES P. MCGOVERN, Massachusetts
TIM HOLDEN, Pennsylvania
NICK LAMPSON, Texas
JOHN ELIAS BALDACCI, Maine
MARION BERRY, Arkansas
RONNIE SHOWS, Mississippi
BRIAN BAIRD, Washington
SHELLEY BERKLEY, Nevada

SUBCOMMITTEE ON OVERSIGHT, INVESTIGATIONS AND EMERGENCY MANAGEMENT

TILLIE K. FOWLER, Florida, *Chairman*

LEE TERRY., Nebraska *Vice Chairman*
JOHN DOOLITTLE, California
JOHNNY ISAKSON, Georgia
BUD SHUSTER, Pennsylvania *(ex officio)*

JAMES A. TRAFICANT, JR., Ohio
JERROLD NADLER, New York
SHELLEY BERKLEY, Nevada
JAMES L. OBERSTAR, Minnesota *(ex officio)*

PREPAREDNESS AGAINST TERRORIST ATTACKS

Wednesday, June 9, 1999

HOUSE OF REPRESENTATIVES
SUBCOMMITTEE ON OVERSIGHT, INVESTIGATIONS AND EMERGENCY MANAGEMENT
Committee on Transportation and Infrastructure
WASHINGTON, D.C.

The subcommittee met, pursuant to call, at 2:08 p.m., in room 2253, Rayburn House Office Building, Hon. Tillie K. Fowler [chairwoman of the subcommittee] presiding.

Mrs. FOWLER. Would the meeting please come to order. I want to thank all of you for being here this afternoon and we are going to have an interesting hearing. Today, our subcommittee is going to examine the Federal Government's programs that are designed to assist State and local emergency officials in preparing for a terrorist attack involving nuclear, biological, and chemical weapons.

The efficiency and effectiveness of Federal preparedness programs are not abstract issues if you receive a call that your child's school is being evacuated and you count on firemen, police, and paramedics to save and protect your child before you can get there.

If you are confronted by the news of a huge explosion downtown, as one of our witnesses today has been, at your husband or wife's office, the last thing you need to worry about is whether the firemen or police are prepared to deal with that situation. We all automatically assume that they are prepared. Unfortunately, when you're faced with a weapon of mass destruction, not all first responders are ready today.

The Congress recognized this some years ago and has passed a number of laws and appropriated funds to bring Federal resources to bear on this problem. A large number of agencies have gotten involved in this effort and are experiencing dramatic increases in funding. For an example, this chart that's over here—and I don't know how well you can see it, but you might want to take a little look at it later—shows how the budgets of just three agencies, the FBI, Health and Human Services, and the Office of Justice programs have increased. That's a pretty dramatic increase in funding levels.

[Chart]

Spending on Preparing for and Responding to Terrorist Acts

Agency	FY 1999 Spending ($ millions)
National Security Community (includes DOD, intelligence agencies)	592
Department of Justice	291
Department of Health and Human Services	130
Department of Energy	98
Department of Treasury	81
FEMA	16
Other Agencies	25
Total	**$1,233**

Source: Office of Management and Budget

The President's fiscal year 2000 budget request of $10 billion for unclassified programs combating terrorism is a $3 billion increase over fiscal year 1999 and a 43 percent increase over what was spent in 1996. Most of this increase has gone to the Department of Defense.

We're about to hear witnesses testify that there are serious problems in these Federal programs. These are not problems of quality. Federal agencies should be commended for assembling world-class training programs and response personnel. The problem is that there are a multitude of fragmented and independent Federal programs that are confusing the very local emergency officials that they are intended to help.

Witnesses will tell us that training and response teams appear unnecessarily redundant and inefficient. As one local official rhetorically asked after participating in yet another Federal anti-terrorism program"how many ways can you bake the same chicken?" These problems have been brewing out there for some time and thanks to the efforts of Representatives Chris Shays and Ike Skelton they are being brought to light.

The Administration is also aware of these problems and has created a new office in the Department of Justice to try and address some of them. We believe this is a step in the right direction. But I will tell you that I share the view of some of our witnesses that the ability of this new office to rein in diverse entrepreneurial Federal programs seems extremely limited. I have strong doubts that simply getting everyone in the same room to talk will make all these problems go away.

I also question why FEMA, who is the lead agency for Federal preparedness and response activities, is not assuming a stronger leadership role. Since this subcommittee has both oversight and legislative authority over Federal emergency management issues, we are uniquely situated to identify and legislate on these problems. We will be taking a closer look at these issues today and in the coming months.

State and local governments deserve the best support that we can give them. We need to make sure that these resources are translated into the best trained, best equipped and best supported fire, medical, and law enforcement officers in the world. And this should not be true in just Washington D.C. or Los Angeles or New York, but anywhere in the United States where we may face a weapon of mass destruction.

I look forward to hearing the testimony from all of our witnesses and working together on achieving this goal. I would like now to turn to my ranking member and good friend, Mr. Traficant, for his opening statement.

Mr. TRAFICANT. Thank you, Madam Chairwoman, and I congratulate you for calling this hearing on the state of the domestic preparedness against terrorist attack involving nuclear, chemical, or biological weapons. I'm going to deviate a little bit from my prepared text, Madam Chairwoman, because later today or tomorrow I will have an amendment on the House floor.

The amendment is very straightforward. It would allow the use of military personnel on our borders to ensure that we might be able to stop penetration of terrorist threats and narcotics.

Nearly all terrorists finance their nefarious activities with the lucrative narcotics trade. What amazes me is a recent report that came out, Madam Chairwoman, that only three of every 100 trucks crossing the Mexican border is even inspected. Experts are saying that nuclear weapons and devices could be smuggled across our border while we are debating a bill that will ship 7,000 American soldiers to guard borders in Europe. Our borders are wide open.

Now, that's not the purpose of our meeting here. But everybody here assembled is familiar with the tragedy of Oklahoma City.

In fact, Madam Chairwoman, I'm going to be asking for your support later in this session when our committee marks up the Traficant Bill H.R. 809 that will make long overdue changes in the Federal protective services. But to underscore the importance of your hearing here, at the time of the tragedy at Oklahoma City, there was only one contract security guard on duty covering three Federal buildings. This is a disgrace.

I believe the Chairwoman is exactly correct when she said someone has to be the bottom line and the buck must stop there. And it's FEMA who has the preparedness responsibility and coordination. I don't see any coordination. Very few agencies really work together. In fact, Madam Chairwoman, there's a lot of competition for some reason among some agencies and not enough communication and camaraderie.

With that, I would ask that my total statement be incorporated into the record and I look forward to interacting with these panel members.

Mrs. FOWLER. Thank you, Mr. Traficant. I appreciate those comments.

Mr. Terry, do you want to make an opening statement?

Mr. TERRY. For the sake of saving time, I do have a statement that I will submit to you for the record; but I will make one quick comment. As an 8-year city council member in Omaha, Nebraska, which, by the way, was the runner-up city for McVeigh, in sitting down and talking to our firemen and our police, one of their major frustrations is lack of support and lack of training in case Omaha, Nebraska is a target for a terrorist attack.

And as McVeigh proved to the world, these are real threats, not only as you said, Madam Chairwoman, for New York and Los Angeles, but for the Oklahoma Citys and the Omahas of the world. So this is important.

Like I said, I have sat down and talked to our firemen, and they're frustrated that they aren't receiving the type of support and the training that they feel is necessary as first responders. What we all have to keep in mind is that whenever one of these attacks is pulled off in our country, the first people there are our local firemen and our local police officers. And those are the folks that need to be trained.

So the competition between agencies at the Federal level is just frustrating the ability to help folks in a real emergency at the local level.

So I look forward to these hearings. And hopefully we can find a solution to this problem.

Mrs. FOWLER. Thank you, Mr. Terry. Mr. Isakson, do you have an opening comment you want to make?

Mr. ISAKSON. Thank you, Madam Chairperson. Only a couple of observations. One is that I don't think there's any question that the major threat to our people in the 21st century is from terrorism and from terrorism on our own soil. We have seen in the World Trade Center, we saw in Oklahoma City those two instances which were warnings. There is no question that it can and there is, unfortunately, no question that it will happen again.

In preparing to listen today to the testimony by studying what was provided to us a few days earlier, it is apparent that there does have to be a place where the buck stops. There does have to be a line of accountability. And wherever there is duplication it needs to be streamlined. Regardless of what we do in the Federal Government, the first line of defense in our home communities are the firemen and the law enforcement officers.

In fact, if we look back on the two terrorist attacks most prominent in America in recent years, it was a law enforcement officer who saw an expired license tag on Mr. McVeigh's car that actually arrested him. And the World Trade Center, it's only because they attempted to go back and get a refund from the U-Haul agency on the van they rented that we were able to secure those and prosecute those individuals.

Again, local law enforcement in both those cases played the primary role. We owe it to our citizens for them to have as much comfort as they can possibly have that we are prepared to respond quickly, and we're prepared at the Federal level to support those in the local level who ultimately will be the first line of defense and the first response in any terrorist attack.

Mrs. FOWLER. Thank you, Mr. Isakson. I believe that is all of our members here. I have a unanimous consent request to allow members who are not members of the subcommittee to sit in today and these are Representative Chris Shays who is Chairman of the Subcommittee on National Security, Veterans Affairs, and International Relations of the Committee on Government Reform; and Representative Ike Skelton who has been closely following this subject for a number of years. I ask unanimous consent to allow Mr. Shays and Mr. Skelton to participate in today's hearing. Any objection?

Without objection it is so ordered.

Now, I would like to—when we had our first subcommittee hearing—just remind people that what we had decided in the subcommittee that we will adhere to the 5-minute rule today which will apply not only to witnesses but also to the Members.

And what we'll do, there aren't too many members, we can go back around, but at least make sure everybody gets a chance to ask some of their questions and we'll go around again for their other questions also.

I would also like to tell both of the first witnesses and all the witnesses that your testimony and that of all the other witnesses will be included in its entirety in the record. As you speak, if you could summarize your testimony that would be very helpful to us as we go through the hearing today.

Before we proceed with testimony, we will swear in the witnesses. This has been the practice of this Subcommittee going back to its earliest days. The Chair notes that this is not a hostile hearing. Nevertheless, copies of committee rules spelling out your rights and the limitations on the power of the subcommittee are available. The Chair also notes that you're entitled to be advised by counsel during your testimony if you so desire.

I would like now to introduce our first panel, Mr. Mark Gebicke who is director of national security and preparedness issues for the General Accounting Office. I believe you have two other members from the General Accounting Office with you today, Mr. Gebicke. If you would like to introduce them and then I'll swear all three of you in.

Mr. GEBICKE. To my left is Richard McGeary and to my right is Davi D'Agostino.

Mrs. FOWLER. If you would please rise and raise your right hand.

[Witnesses sworn.]

Mrs. FOWLER. Thank you so much. You can be seated.

Mr. Gebicke, if you would proceed.

TESTIMONY OF MARK E. GEBICKE, DIRECTOR, NATIONAL SECURITY AND PREPAREDNESS ISSUES, NATIONAL SECURITY AND INTERNATIONAL AFFAIRS DIVISION, U.S. GENERAL ACCOUNTING OFFICE

Mr. GEBICKE. Thank you, Madam Chairwoman. We're pleased to be here to continue our dialogue with members of the Congress about this important issue. And what I'll do within my 5 minutes is to share some observations in just three key areas:

One, the threat of a chemical, biological, radiological or nuclear incident taking place in the United States. Secondly, the rapid Federal growth of both training activities as well as response activities on the part of the Federal sector that you alluded to in your opening remarks. And finally, some of the steps that the executive branch has recently taken and most importantly some steps, some very important fundamental steps, that we think are still missing from the whole program.

Let's turn first to the threat of terrorism in the United States. Now, the U.S. intelligence agencies are continuously monitoring threats both from inside the country as well as from outside the country. And what they tell us is that terrorists are less likely to use chemical and biological and nuclear weapons than they are to use explosives and firearms for a couple reasons.

Number one, it's much more difficult to actually procure and produce and disseminate chemical weapons and biological weapons than it is the more conventional types of weapons. Also those weapons are much more unpredictable. You just don't know what's going to happen when you release them.

Now, the intelligence community also tells us that over the next decade or so threats from these types of weapons—chemical, biological and nuclear—might be increasing. So it's something we need to watch and something that is evolving.

Unfortunately, we find also that there are a lot of conflicting statements made in public forums, particularly with regard to chemical and biological weapons and with regard specifically to the ease or difficulty with which they can be delivered.

On the one hand, some people have suggested that it's relatively easy to produce effective chemical or biological weapons at home using recipes off the Internet. Others suggest that it's far more difficult than that.

In addition, there appears to be a disconnect between the intelligence agencies' judgments and the focus of at least one program that we're aware of. And we can get into that a little bit later.

My second point—and you alluded to it in the charts that you put up at the very beginning of the hearing—significant increase in the budget requests—$6.5 billion in 1998, $10 billion budgeted or proposed for the upcoming fiscal year. The Federal Government has created similar programs, both to provide training to the State and local responders as well as to actually provide a Federal response capability.

In our statement we've laid out some details of those programs and also the Federal responses that are involved. It goes without saying that the more players you have involved, the more difficult it becomes to coordinate and integrate the response.

The final point is to acknowledge and recognize some steps that the executive branch has taken. These are very positive steps and they shouldn't be minimized. In the last couple of years OMB put together all of the budgets for the Federal agencies in combating terrorism. That's the first time that had been done.

We now have that information. We know who the players are, and we know to some extent the resources from the taxpayer are going into this program. The attorney general has issued a classified 5-year interagency plan on counterterrorism and technology. The plan was very well done.

And the Attorney General has also planned to establish a National Domestic Preparedness Office at the FBI, which will hopefully coordinate better the various activities that are taking place and make it easier for the State and local responders to understand what's available to them.

Now, having said that, and having talked about the positive nature of these steps, there are still three opportunities where we see that things can be improved. One is that we still don't have a government-wide strategy, and we don't have a defined end-state.

If we had an end-state I think we could answer some of these following questions, if not all of them. What's the end game for all of the nations' investment and programs? How are we going to get there? What's our road map? How are we going to know when we have achieved what it is we want to achieve? Is the desired end-state for every city and every local entity to have a full capability to respond to terrorist threat? Is that reasonable?

Secondly, I think we need to establish firm program requirements. And these have to be based on threat and risk assessments. They have to be. If we put these in place, these assessments and these inputs, we can make decisions then on programs, requirements, and investments. And we can make sure that those investments are geared towards the more likely threats.

Right now it appears as if we're trying to prepare for the most catastrophic threats, which may not necessarily be the most likely. And then finally—and I'll close in just a minute—we really need a comprehensive inventory of all the assets that are involved in combating terrorism at the Federal, State, and local levels. Only when we know what's out there and what the requirements are, can we figure out what we need to add.

Now, if we put these three program elements—very fundamental elements—in place, I think we'll be in a better position to know whether or not we have established the right programs funded in the right relative amounts. And then I think we're on the right track.

In closing, I would just say that we are not here to say that this is not a good investment for the Federal Government. It obviously is very critical—it is a good investment. What we are saying is we think we can make it a more effective investment through better focus and better direction. Thank you, Madam Chairman.

Mrs. FOWLER. Thank you, Mr. Gebicke.

The general accounting office always does such a good job. We always depend on you to help us wade through the myriad of details with these programs and help us come to some decisions. I think the questions you have raised are very good, valued ones. We share your concerns. We'll be looking at some of these today and in the coming months.

One point I would like to make from something you said in the beginning—and this is maybe because I'm on the Armed Services Committee—I have great respect for our intelligence committee, but they were way off base by about 8 years on the missile capabilities of some of the rogue countries around this world.

So if they're that off base on whether we're going to be facing terrorist attacks using any of these weapons of mass destruction, I think we better be getting ready now rather than waiting years down the road.

We have seen that with missile defense. I'm concerned with seeing what happened in Japan a couple years ago. It's very easy nowadays to acquire a weapon and use it if someone so chose. So I have a little dispute with our intelligence committee evaluation that you were quoting there.

As you pointed out in your testimony, there are about 40 different Federal agencies that are involved right now in combating terrorism. There's more and more money being spent each year on this and we all agree we need to spend the money to do it right.

In reading through your testimony, do you see who in the Federal bureaucracy is in charge of all this? Have you been able to identify any one entity that is in charge?

Mr. GEBICKE. Well, probably the closest person we have to being in charge is at the National Security Council, the National Coordinator for terrorism. But obviously there are some inherent limitations in that particular function. The coordinator does not have the role to direct the agencies to do things. It is more of a coordinating role, as the name implies. It's an integrating role. It's a role of persuasion.

And also I guess from your perspective I would have some concerns as to whether or not that individual also would report to the Congress. So if that person—I think a true coordinator or person in charge would be somebody who would come up and sit before you and discuss the issues. I guess for the reasons that I just laid out it's probably not truly somebody 'in charge.' But that's the closest thing that we see at this point in time.

Now, there is something that I will add that we are just embarking on right now. We are getting ready to do a review of how other countries—specifically countries who have dealt with terrorism issues for a long period of time—are organized to address this issue in their countries. And right now in our scope we're thinking about going to Israel, the United Kingdom, France, and Germany. It will be awhile before we have the results, but we hope from that study to bring back some good practices, if you will, that can be considered and possibly determine if they might be applicable to our situation here in this country.

Mrs. FOWLER. That's interesting. I look forward to reviewing the results of that study. I know if you have been doing some of your work, you have been talking to State and local officials around this country. Do you get any specific feedback as to what Federal entity they think should be in charge of this or who they should be looking to be coordinating this?

Mr. GEBICKE. No, I don't think we have heard exactly who should be in charge. What we hear most often is there are too many of them. It's just too confusing and too hard to deal with.

Mrs. FOWLER. I agree. Based on your research and people you had been meeting with, what do the people you have talked to think about FEMA's training programs? Is there any criticism regarding the way these training programs of FEMA's have been implemented?

Mr. GEBICKE. We really haven't looked at the FEMA training program in detail so I can't comment on that. We are in the process of starting two other assignments; and if you would like, I could go into those just briefly.

Mrs.FOWLER. OK.

Mr. GEBICKE. We are looking at all the various training programs that the Federal Government is providing. So we're going to be doing a fairly comprehensive inventory trying to determine where there's overlap and where there are gaps in those training programs. In another assignment we're also going to be looking at the duplication and overlap, if it exists, in the Federal response assets. We're going to look at both.

Mrs. FOWLER. I think the members of the subcommittee will see, as we work our way through this today, there's a great need for those studies and the answer to those questions that we can take and move forward with.

If the Department of Defense had taken a different approach in their training programs than just going directly to the cities, could their programs have covered a greater proportion of the population than they're currently covering?

Mr. GEBICKE. Yes, ma'am. That's very possible. There are other approaches that we identified in our work that DOD could have considered. One could have been a county-wide approach and instead of covering 22 percent of the population by using a city approach, they could have doubled that and picked up 44 percent.

And in addition, if they had taken an approach directed at the metropolitan statistical areas, it's also possible that they could have picked up up to 64 percent of the total population. But more importantly than even expanding the message, what they told us at the State and local level is had the approach to the course been organized differently to recognize the mutual aid agreements that exist within each State and locality, it might have worked better.

To give you an example, if it had come to the Washington metropolitan area instead of to Washington, D.C., we could have included Fairfax, Arlington, the City of Alexandria, PG County, Montgomery County, as well as the District of Columbia. And that's basically the mutual aid agreement that exists in this particular section—in our corner of the world.

Mrs. FOWLER. So as DOD went around the country implementing their training program they really just ignored those mutual aid agreements.

Mr. GEBICKE. They went to a city-by-city arrangement. But we understand—and we haven't been able to verify this—that they have tried to include some of the other entities in the more recent activities. So I think they've recognized it the problem.

Mrs. FOWLER. Good. Just one last question and comment. You referred to the Attorney General's 5-year plan on technology and counter terrorism. My question was does it solve the problems of coordination and focus that we are concerned with today? This is a secret report so you have to have secret clearance to read it. I sent a staffer over to read it since I didn't have the time to go see it and read it, and we couldn't get it for him.

I would like to hear your opinion, then I'll tell you what ours is.

Mr. GEBICKE. I hope we agree. My assessment would be that this plan gave us a lot of information that didn't exist. I don't want to minimize by any stretch of the imagination the preparation of the plan. Fifteen different agencies were involved. It was a very time-consuming, laborious effort. However, the plan doesn't link programs to assets or resources.

And it really doesn't solve the three major problems that I laid out for you in my opening remarks. It doesn't give us an overall strategy. It doesn't give us an end game. It doesn't figure out what the requirements are. We don't have an inventory.

So the short answer is no, ma'am, it's a step in the right direction; but it doesn't get us there.

Mrs. FOWLER. We're in total agreement. I was most disappointed because it's sort of a plan to do a plan. Yet it was supposed to be laying out a 5-year plan for how we deal with this, but it doesn't. So we've still got a lot of work cut out for us as a result.

I want to recommend to the subcommittee they make sure they get the information in that too. I will turn to the ranking member, Mr. Traficant, for his questions.

Mr. TRAFICANT. Thank you, Madam Chairwoman. I ask that the answers be very brief, and I will submit most of my questions in writing and ask unanimous consent that they be answered in total and submitted for the record.

But I do have a couple points in line of what you talked about. You know, we learned about the collapse of the Soviet Union on CNN, the fall of the Berlin Wall on CNN; we learned about the invasion of Kuwait by Sadam on CNN. And I made a statement once, why do we fund the CIA? Why don't we hire CNN?

So I'm a little bit more skeptical on this terrorist issue. Madam Chairwoman did address the main question I had was who's in charge. I don't think we really know. But there are several concerns I have, one is sharing of information, dissemination of intelligence-gathering with mutual flow between entities who need such matters.

So you've done this review, and you've taken a snapshot look at it—is it Gebicke or Gebicke?

Mr. GEBICKE. Gebicke.

Mr. TRAFICANT. Mark.

Mr. GEBICKE. Right.

Mr. TRAFICANT. And I'm one that you know has a lot of questions about a lot of government agencies. And this isn't to patronize you, and I think that GAO has been one of the more reliable agencies Congress has; and I learned to respect your judgment and, in fact, ask for it quite a bit.

In your review, just briefly, does anybody do anything of any measurable performance evaluation? Yes or no.

Mr. GEBICKE. I would have to say yes, given the way you phrased the question.

Mr. TRAFICANT. Then let's——

Mr. GEBICKE. Can I expand just 30 seconds?

Mr. TRAFICANT. Briefly. I only have 5 minutes. You know the bell.

Mr. GEBICKE. I'm thinking of the Government Performance and Results Act. We have seen some strides in the form of strategic plans, and we have seen some annual performance plans on the part of the agencies. I don't get personally involved in reviewing those, but I know we have seen some progress in that regard.

Mr. TRAFICANT. Are we getting our money's worth?

Mr. GEBICKE. Are we talking about in this particular area?

Mr. TRAFICANT. Yes.

Mr. GEBICKE. It's a difficult question. I don't think we know yet where all our money is going. I don't think we know what all of our capabilities are. And I think we have to leverage existing capabilities. It's not just a Federal problem. As you all mentioned the first people that respond are at the State and local level. So we've got to integrate our resources, and we've got to do what's in the best interest of the public.

Mr. TRAFICANT. We have the WMD, the NLD, the DOD, the WMD, the FEMA, NDPO, DDT. I don't mean to make light of it, but it seems we have so many people in this field and we just basically get news. So you have done a review, and I think GAO has been pretty reliable.

Before I get to one significant question, just, if you would, on your own, on a 1 to 10, gauge it. What is the risk to America on terrorist threats pursuant to what your review has uncovered on a 1 to 10, if you can do that. I know it's tough.

Mr. GEBICKE. Ten being the most catastrophic?

Mr. TRAFICANT. Yes.

Mr. GEBICKE. It really depends on a formal risk assessment. I mean, if you're talking about a risk of a massive conventional bomb, the risk is going to be much higher than what it would be to experience a catastrophic incident involving chemical or biological weapons. Nuclear would be really, really low.

So I would have to say for conventional and as the size goes up and the level of the catastrophe increases, the risk probably drops down because the more materials you would have to assemble to make it a catastrophic event. So I don't know what the scale would be, but I would have to put risk of experiencing conventional attacks higher than the chemical, then the biological and then the nuclear.

Mr. TRAFICANT. I guess this would be my last question: Is it that those who are assessing the risk—is it because they're not sharing the information that causes us greater exposure? Or is it that they're sharing, but we're not appropriately gathering the information to pursue the right planning mechanisms? And that's in line with the Chairwoman's question which was my very first, who's in charge?

See, at some point you have all these different entities, and everybody thinks that somebody else is going to do it while anybody is supposed to have done it, but nobody gets it done. And we look at the Oklahoma City case. There were a lot of anomalies there: anniversary dates of Waco, certain other unusual elements that were there. It seemed that nobody really shared any of that technical data.

So, again, is it the sharing or is it we're not gathering it properly or that after we do it, everybody expects somebody else to carry it out and let the locals who are to be trained and coordinated basically carry the load?

Mr. GEBICKE. I can't give you a good answer on that because we really haven't looked at the intelligence community specifically and how they go about gathering their information and combining it. We do have a report that is going to come out later this summer that's going to take a look at how or if the intelligence agencies have pulled together their information in terms of what the threat is. And that might answer your question when we complete that work.

Mr. TRAFICANT. Is the money we're giving for this effort getting down to the fire departments and police departments in reasonable training? Just yes or no.

Mr. GEBICKE. In the case of the DOD program, we got very favorable responses at the State and local level as to the quality of the training courses. They thought the course was very good. There is also then $300,000 per city in equipment provided to each city. This enables them to buy protection gear, decontamination equipment, identification equipment, and also training aids.

We haven't looked at the others in any great detail at this point in time.

Mr. TRAFICANT. Thank you, Mark. My other four questions will be submitted for answer by writing.

Mrs. FOWLER. Thank you. I want to welcome our other colleague, Congressman Berkley, who has joined us.

Mr. TERRY. Thank you. Mark, first an editorial comment. This should not just be a Federal program that we have to leverage, at least in towns like Omaha. It's only a mere 600,000 folks.

We are the only source right now looking forward to the DOD-type programs. What we're looking for at the city level is the complementary-type programs. There's a feeling that Orange County and New York may have the capabilities as localities to have the level of training where, in fact, they may be able to come in and better train than some of the DOD folks. But that's rare.

You know that's inherent to those large communities, whereas the Omaha, Nebraskas, the Oklahoma Citys look outside of our boundaries for that level of training. But right now we look within ourselves for not only all of the funding for it, but the training, our hazmat training.

While I think we would be very good if a gasoline tanker turned over within the city limits, I'm not sure how they're trained to react if there was a biological issue. And that's where we look towards the Federal Government or the DOD that would have that level of expertise. So I think when we're doing 100 percent of the work right now, we're just looking for some help.

Mr. GEBICKE. Absolutely.

Mr. TERRY. I want to make that clear. If I'm still in my capacity as a city council, I'm not saying Federal Government take over something for us.

Mr. GEBICKE. 'Give us a hand; we can't do it all ourselves.'

Mr. TERRY. Give us a hand. While we're very confident in our own abilities, we may be dealing with something that we haven't dealt with before and need that extra training. We may need some extra help financially and with training with special type of equipment that's necessary for special types of problems, that aren't inherent in the day-to-day tasks of a firefighter, EMT or police or hazmat operation.

So I just want to go on record and make that statement. I don't want there to be any feeling that cities just want the Federal Government to take over these programs. That's not what we want or what they want.

Mr. GEBICKE. If I could comment. That is the comment we heard also at large cities. There's some nervousness among officials at the State and local governments that maybe if there is an incident that the Federal assets will come in right away. And they would like to deal with those problems to the extent they can. If they feel they're overwhelmed or they don't have the capability, then they will ask for assistance. It's a similar theme in the larger city as it is in maybe the smaller or midsize city.

Mr. TERRY. That's good to hear. Here's the one question, and it's certainly carrying on the themes that we heard from our opening statements and the two questions. The Department of Justice's requested for fiscal year 2000, 162 million for training and equipment to localities, the first responders in prior legislation passed by the 104th Congress—funds were allocated to the Department of Defense for this same purpose as well as the FBI, Department of Health and Human Services, and other departments, as Mr. Traficant pointed out.

National response teams, according to legislation, are to be established through HHS. So my confusion here is why so many are tasked with the same tasks? Why is the Department of Justice currently funding this program when part of the responsibility is with HHS? And this just boils down to the same frustrating question that's already been posed to you, who should be in charge of this?

Mr. GEBICKE. I don't know if I could answer it any better than my attempt earlier. You could probably make an argument at a number of different locations throughout the Federal Government to have one agency in charge. The closest thing we have right now is the Coordinator for terrorism, National Security Council.

Mr. TERRY. Very good.

Mrs. FOWLER. Thank you, Mr. Terry. Mr. Isakson.

Mr. ISAKSON. Mr. Gebicke, I was reading in your prepared testimony a statement that said on the surface it appears to us that there's potential for duplication and overlap among these programs. You were referring to training.

In the materials we were supplied with, there's a report from May which listed the number of training programs in the country other than those in DOD. I happened to count them up, and there were six agencies that had 85 different training programs. And interestingly enough, in three different cases there are programs with exactly the same title produced by two different agencies.

And so in light of your observation that it appears there may be duplication, it appears to me there probably is duplication. You made the statement we have a coordinator at the National Security Council. Is that right? Is there any coordination whatsoever, or are these departments operating independently on the development of these training programs?

Mr. GEBICKE. Well, in the case of DOD, when they developed their training program, they basically developed the program on their own. Programs were also being developed or in existence at the Department of Justice and FEMA at the time. Program material was offered to the DOD, but DOD opted to develop its own program, to give one example.

But what we want to do on our current assignment is, as I mentioned earlier, is—I agree with you on the surface they look like duplications—take a closer look at the courses. We need to maybe sit through some of those courses, have some of my staff sit through those courses, make sure they're being targeted to the same audiences, that is, police, fire, hazmat; and then we'll be in a much better position to come back to you and really point out the exact duplication or overlap of the training programs.

My only hope and my only concern would be it wouldn't be too late at that point in time because they proliferated to such great extent.

Mr. ISAKSON. Well, first of all, I think your observation is correct. I think your recommendation is correct. You don't want to pass judgment and say it's obvious. But it's apparent—there's a chart here somewhere with regard to DOD and FEMA's training courses, which I have a copy of. I'll show you this chart, the staff evaluated the DOD courses and the FEMA courses by category, and it looks kind of like a fire drill of some type.

Mr. GEBICKE. I don't think you're going to be able to explain that in 5 minutes.

Mr. ISAKSON. I used to call it a Chinese fire drill, but I'm reluctant to do that now. If you look at the categories on the left and right under FEMA courses and then you see the lines, in each and every case the courses address the same subjects but may be under different title.

You just said that you did not believe that DOD had utilized the resources of any other department's studies or course of training programs. So it would seem to me to be extremely important that GAO do follow through and test these training programs and if there is duplication, find out some way to coordinate it so we're not recreating the wheel.

I would make one other comment if I could. Mr. Terry, Representative Terry, made an outstanding comment. I agree with you that the likelihood of a nuclear, biological, or chemical incident in this country is far less likely than a conventional incident.

We need to be aware that we have already had in this country incidences with conventional weapons in the World Trade Center, and in the case of the use of common fertilizer in Oklahoma City. We must remember they used sarin gas in Tokyo in a subway. And I think, although the rating would not get a 10, I think your answer is honest.

I think it would be less than responsible Congress of we didn't recognize that it is more than likely that in the first decade of the 21st century, the potential for the next type of terrorist act to involve an attempt to use chemical weapons which are very difficult to deploy, or chemical acts. Nuclear, I disagree is probably more unlikely, but it is likely there will be an attempt—hopefully there wouldn't be an incident—and Mr. Terry's comments are so good. Our local law enforcement, fire, police, EPA, transportation people, I think are very, very good at chemical spills on highways.

I think they have. In Atlanta we had an unfortunate terrorist incident during the Olympics. The response was totally local, even though we had a significant amount of Federal assistance in the area. When it comes to the nuclear weapons, biological agents, and chemical agents, I think we must look to the Federal Government to be the first line of information, strategy, and attack. Because the resources to deal with those types of incidents which go far beyond a localized explosion are incredible.

If we have six agencies writing 85 reports to the Federal Government and we have 50 States and the District of Columbia writing training programs on chemical spills, we'll have a disaster on our hands because the right hand doesn't know what the left hand is doing.

So I appreciate your report very much. I think your observation is correct that there's too much duplication. I really think we need to do what Mr. Terry said. That is, focus on where we in the Federal Government can deliver the first and the most rapid response in the more unlikely but all-too-probable nuclear, chemical, or biological incident so our local people know they can count on us. So we're prepared to deploy.

Mr. GEBICKE. Can I make one comment? And I think it's an obvious comment to all of us here in the room and that is we're not going to protect ourselves against everything. So we have really got to separate what's conceivable, because I think almost anything is conceivable, particularly if we're talking about maybe the first decade of the next century, from what is likely. And I think we need to protect ourselves against what is likely, because I don't think we can protect ourselves against everything that could happen.

Mr. ISAKSON. No, but we must remember the budget that we're considering right now is the budget of the first decade of the next century.

Mr. GEBICKE. Exactly.

Mrs. FOWLER. Thank you, Mr. Isakson.

Mrs. Berkley, did you have a statement or questions?

Ms. BERKLEY. Well, I did have a statement. About a hundred years ago, when I first graduated law school and went back to Las Vegas to practice law, I was appointed deputy director of the State commerce department. And in that capacity I was charged with researching what local entities and State entities would be available in case of a disaster, either natural or man made. And it was quite an eye-opening experience, one, because there were so many different local and State agencies involved, not to mention Federal agencies.

But the second thing is that nobody was prepared to do anything in spite of the tremendous number of agencies involved if, in fact, we had a serious life-threatening tragedy throughout the State of Nevada. So this is an issue that I dealt with about 25 years ago, and that's why I was very interested in coming here today and reading your testimony, which I will review, and then, if you don't mind, submit questions for more thorough answers to the questions that I have.

I do believe that this is a very dangerous world and there will be a time, I fear some time in the beginning of the 21st century, that we might have a catastrophic tragedy somewhere in our country. We need to be prepared for that eventuality. I am concerned about it, interested in it, and anxious to work together to come up with a meaningful way of dealing with this. Thank you.

Mr. GEBICKE. We'll be glad to respond to your questions.

Mrs. FOWLER. Thank you.

Do any members of the subcommittee have any follow-up questions that they would like to address to GAO at this time?

I want to thank you, Mr. Gebicke, and your associates for your work. We look forward to these studies that you are in the process of beginning because this is information we need, as you say, sooner rather than later. As money is being spent and Agency's expenses climb, we have got to work out a better coordinated effort on this so that the money is being spent wisely and efficiently. We look forward to continuing to work with you. This is an issue that we are going to continue to follow in this subcommittee in the coming months. So we'll look forward to working with you on it again. Thank you.

Mr. GEBICKE. We look forward to working with you. Thank you very much.

Mrs. FOWLER. Thank you. I'd like now to call our next panel, if they could come forward.

We have Chief John Eversole and Councilwoman Ann Simank. If you don't mind, we will take a brief recess, so our members could go vote and come right back. We only have one vote. Then we will come back and reconvene the hearing. We are recessed for the time of this vote. Thank you so much. We will be right back.

[Recess.]

Mrs. FOWLER. Would the subcommittee come to order, please. I would like to welcome before us our second panel, Ms. Ann Simank, who is a member of the city council in Oklahoma City. She's here today on behalf of the National League of Cities. I also welcome Chief John Eversole, who is a hazardous materials coordinator for the Chicago Fire Department. He's the chair of the hazardous materials committee of the International Association of Fire Chiefs.

I want to welcome you both; and before we begin, I would like to ask if you heard me earlier, because our practice is to swear in our witnesses, if you could both stand and raise your right hand to take the oath.

[Witnesses sworn.]

Mrs. FOWLER. Thank you. Please be seated.

I think that we'll just go ahead. If you don't mind, we'll start with Councilwoman Simank and then go to Chief Eversole.

TESTIMONY OF ANN SIMANK, OKLAHOMA CITY COUNCIL MEMBER, ON BEHALF OF THE NATIONAL LEAGUE OF CITIES; AND CHIEF JOHN M. EVERSOLE, CHICAGO FIRE DEPARTMENT, AND CHAIR, HAZARDOUS MATERIALS COMMITTEE, INTERNATIONAL ASSOCIATION OF FIRE CHIEFS

Ms. SIMANK. Thank you, honorable Chairwoman. Good afternoon. I am very pleased to be with you. Before sharing with you the positions of the National League of Cities on domestic terrorism and related issues, I would like to thank the subcommittee for providing this forum of discussion on the importance of domestic preparedness in the event of a terrorist attack.

I am Ann Simank, council member from Oklahoma City, Oklahoma, and I am testifying, as you heard, for the National League of Cities. I have been a member of the National League of Cities public safety and crime prevention policy committee for the past 3 years. This NLC policy committee has given considerable attention recently to the many issues that we've talked about here today.

On April the 19th, 1995, at 9:02 a.m., Oklahoma City, my hometown, was devastated by a horrendous terrorist attack, the bombing of the Alfred P. Murrah Federal building. That fateful morning 168 lives were lost, thousands were seriously injured, and over 30 children became orphans, and approximately 300 businesses were totally devastated and put out of business that minute. Many lives were literally in chaos in our city.

This senseless act of terror shattered my city, all in a matter of just seconds. Were we prepared? I don't know if cities and towns in America can ever be fully prepared, but today, I want to share a little bit about my experience with you in Oklahoma City.

Our city had been selected for FEMA's integrated emergency management course in Emmitsburg, Maryland, shortly before the bombing. Approximately 1 year before —and all of our department heads attended this training course, as well as two or three key people from each department, and many of our community leaders attended, chamber of commerce people attended, public utility workers attended, nonprofit organizations attended. The training was excellent, and we were fortunate that we were selected as a city to be able to have that training.

However, Oklahoma City faced difficult challenges on the morning of April 19th. This was the largest terroristic act ever seen in America, and to our knowledge no one had ever had to perform recovery and response at the same time while dealing with a crime scene.

Immediately after the bombing occurred, our mayor and our fire chief and our police chief got together with the local special agent of the FBI, and we just had a decision-making process in a room, and the decisions were made about how Oklahoma City would get out there and proceed. The Federal Government wasn't here yet from Washington.

We decided that the FBI would be in charge of the crime scene, of course, and that the Oklahoma City police would seal the perimeter to keep citizens and other people out of the area because it is a crime scene and that the local Oklahoma City would assist the FBI in the crime scene investigation, and we also decided that the fire department and the fire chief would be in charge of rescue and recovery.

Now, our incident command system was up and operating immediately when these local agreements were made, and it worked well. That morning and all that day we, the local people, were there all alone, and we had made these decisions. We saved lives all day long and into the early hours of the evening the last victim was saved. And we also collected evidence that day.

It was about 15 hours later before the first FEMA team arrived, and so you're going to have cities that sit there 15 hours operating before someone gets there to help them. When the first FEMA team arrived, things changed. Other teams flew in, other agencies and people flew in, and I think because of FEMA's normal recovery process in dealing with hurricanes and earthquakes and so forth they had never dealt with a crime scene. So no one really knew how to deal with all of this. No one knew what the protocol would be. All of those questions came up, like who is in charge here. Turf battles started happening.

In the meantime, Chief Gary Marrs and the fire department continued to do rescue and recovery, thinking there might be someone alive. Fortunately, Oklahoma City had good training. Oklahoma City has always been a city that balances its budget and we had resources, fortunately. Not all cities do, and our incident command center was so strong and we were operating so well that all of the different Federal agencies, law enforcement agencies versus disaster recovery agencies and all of the bickering that was happening could not penetrate our incident command system. So we were able to go ahead and stay in charge of Oklahoma City's incident.

What did we learn? We learned that damage starts at the point of impact. The minute something happens your local responders are the ones that will be there. They are the ones that need to know how to take care of the damage, and they're going to be there alone for many, many hours. We learned that any resources that you, you here in Washington, can make available to us out there will be of very much of benefit to us. We need those resources. But we also want you to iron out those complications that surround the resources.

Training is essential since cities and towns are alone, and equipment is essential. Turf issues should not happen. When you're dealing with people's lives and in an incident of this magnitude, a clear line of command and protocol should be established, and the National League of Cities is very much in support of a national strategic plan.

Money should be made available for training and equipment, and transportation needs to be made available. Transportation for the teams —you have all of these teams in Washington that are trained to come help our cities, but many of our USAR teams had to wait long hours on the runway literally with their equipment packed before a military transport could get there to lift them off and take them to Oklahoma City.

Quite frankly, our city has no confidence in military transport to get your teams out to our cities unless something changes here in Washington to get those military transports operating quicker.

We have learned that whether chemical or biological agents are used, whether the next strike is one involving cyberterrorism or whatever method of terrorism, our cities must be prepared for a variety of threats. Large scale evacuations may have to happen, public health emergencies may be involved if it's chemical or bio. Information flow and management is very critical. We need good communication. We need people trained to know how to detect explosives and how to know when they are getting involved with different types of chemical or bio agents.

I know our time constraints today, and so I'd like to go ahead and summarize by letting you hear what the National League of Cities and the 135,000 city members of the National League have as key recommendations for this committee today. We would like to see the development of a national domestic terrorism strategy with clear policies and comprehensive plans to coordinate the roles, responsibilities, and resources of your Federal agencies, and we would like to see training and equipment and resources provided by the Federal Government, especially bringing it out to us regionally, and we do believe in regional training.

Oklahoma City has interlocal agreements, and we have central Oklahoma that should be trained together, and we should all be responding and helping each other.

A policy needs to be made for sharing certain classified information on threats or potential threats of terrorism with local lawful agencies on a need-to-know basis. The Federal Government should include local governments in the Federal planning process and operations relative to issues and their jurisdictions and target scarce Federal resources in localities that have high public profile or private targets.

Designation of a lead Federal agency, something you all have talked about here today, that actually serves as the central coordinator and information clearinghouse on all available Federal programs and resources is critical. Madam Chairperson, we are most appreciative of the Federal Government's attempts to quell this confusion and to set up an area in an agency, the National Domestic Preparedness Office, the NDPO, with the Department of Justice, and although the National League of Cities commends the work of the NDPO thus far, the problem with this latest Federal effort is twofold.

First and foremost, many departments within other government agencies that have been named, the DOD, the FEMA, DHHS, and Justice Department, still maintain that their agency has the lead in coordinating this effort, and this has fostered a Federal power struggle, and it's involving billions of dollars.

I will tell you my police chief—my fire chief, Gary Marrs, told me last week if we could have those billions of dollars out at the local level in the United States we would be doing well.

Secondly, I ask you, what good are any of NDPO's efforts if it cannot receive the funding it needs to implement its objectives? We at the NLC are looking at the objectives of the NDPO, and we want know will they be met clearly and in reality. What message is being sent to State and local first responders when the Federal Government cannot even coordinate their own efforts?

In closing, the National League of Cities hopes that our recommendations will aid in establishing a clear line of authority and help create clear lines of communication between Federal, State and local governments.

Again, the National League of Cities would like to thank you and the subcommittee for holding this hearing to talk about this subject today, and we look forward to working with you in this effort. Thank you very much.

Mrs. FOWLER. Thank you, Councilwoman Simank.

I'd like now to go to Chief Eversole, and then we will have the questions. Chief.

Mr. EVERSOLE. Madam Chairwoman and members of Congress, it's a pleasure and an honor to be here. You sit and you try to think how do you solve all of those problems in 5 minutes, and I certainly don't pretend to know how to do that, but I would like to make a few comments, and you have my written testimony, and I would like to just make some comments.

First of all, earlier I heard a question of who's in charge, and I always say of what? I don't want to direct traffic. That's someone else's job. Certainly everyone should have a place and everybody should know their place and everybody should be in place. The bottom line out there is that local emergency responders do it every day and they do it well.

Now, I would like to speak for just a minute about programs which support local responders. Certainly Nunn-Lugar was a good move, and the Antiterrorism Act was a good move, and the efforts that were made by the National Fire Academy were a good move, and OJP's funding and programs were a good move, and the consortium training was a good move; but it seems that we don't know how to put everything together, and we've heard that here earlier. Just about everybody has said that.

Now, the bottom line, again, is very simply this. You are Washington, and you are the ones that are supposed to put it together, and it just doesn't seem to happen. There is so much confusion and competition between Federal agencies that they are sometimes more interested in what they are doing than what's getting done on the general end of it.

When we look at Federal agency response, they have created the NDPO or at least semicreated it, or wherever. It sits in limbo, and we told Attorney General Janet Reno that we needed one-stop shopping.

If you're confused and asking questions, how do you think we feel? I work for a boss who has one simple rule. He looks at me in the worst times and says, make this work; and it's very difficult to make it work when you're really not sure where to go. So they came up with this concept of NDPO, and I think that it's a good concept, but I certainly think that it needs to get going.

The Federal Government is way too slow for me. I work in a world that demands excellence. 3 to 5 minutes you expect me in front of your house, doing the job, not 6 months, 18 months, 2 years from now. That certainly wouldn't be acceptable to you if I told you I'm sorry, it's Wednesday, we don't do fires on Wednesday. We'll be there next week. Take two aspirins. We'll see you in the morning. None of those things work. You want excellence from us, and we have to deliver that.

We look at the RAID teams. Nobody asked us about the RAID teams. We were told, by the way, here's $50 million. We're going to spend this. Now, maybe that's going to work real well, and I am most willing to give that a try. I had an opportunity during the Democratic national convention to work with the Illinois National Guard, and they did an exceptionally fine job for us; and maybe it will work well, but now we're thinking about creating 44, or whatever that magic number is, light teams.

I'm the kind of guy that before I buy a car, I want to go out, kick the tires and drive it around the block; and we're buying things that we don't even know are going to work. Let's get it on the road and see if it works, and if it works, then we get more. That seems to me like a reasonable thing to do in management.

We have CBIRF, we have the FBI's emergency response team, we have all these people that are very confusing. I don't want somebody in there tugging and pulling on the device, saying, oh, it's my mine, it's yours, it's mine, it's yours. I need to get the job done.

Now, government it seems to keep inventing the wheel and they tell me it's round. I know the wheel is round. I don't want to hear that anymore. I want you to tell me how it can go faster and last longer, and apparently the government doesn't know how to do that for me. They fail me on the local level.

I have got a lot of really good people, I mean people who really care and seem like they want to make this program work. Apparently, none of the telephones connect between Federal agencies. They must not be able to talk to each other on a computer or either never meet in a crowd or a tavern to have a beer and talk things over because they don't know what each other is doing. It's a secret. I went to one meeting, I was told about secret underwear, and I said, excuse me? Oh, yeah, we got secret chemical underwear. We can't tell you about it. OK.

We have a situation where today we have Federal agencies who made a decision about how they were going to certify chemical masks, and OSHA says if you're going to wear a mask in the civilian world, it has to be NIOSH certified, and NIOSH says we don't have a provision to certify for chemical or biological warfare.

So if the bell rings in my fire house and I send somebody out, I'm in violation of NIOSH regulations. Who figured that out in the government? And I don't want to hear it takes 18 months to straighten it out, because if you call me today you're going to be going out and looking for the BRT coming down the street, that's the big red truck; and you're going to look for that truck and you are not going to wait for 18 months. You're going to want 3 minutes. Where are these guys? And we need that equipment and we need those answers and we need them today, not tomorrow, not 6 months from now. We need them today.

We've had five NDC threats in our city. We have successfully handled all five hoaxes. We think we have a good system. We work closely with the local FBI. We work closely with the lab. We have a way to handle it. Understand this: We go out and do it every day, and we handle some relatively bad situations. What we need to do is to enhance what we already have and we know works well.

I have an army. I have 1.3 million firefighters all over this country. We need to make them better, better, not create new things. I don't want to hear about you making that round wheel, folks. Show me how to make my wheel go faster and better.

That's how you talk about—sir, you asked a question, Mr. Congressman, about are we getting our money's worth. No, I'm not getting my money's worth.

I want to see money coming to the people that are at the very tip of the arrow, and I want to see enough money come that you know that when you call they are going to show up with that big red truck and they are going to do the job, that our police officers can do the job, that our EMS people can do the job, that your local responders are going to do it in 3 to 5 minutes, they're going to be there doing it. If we have a chance to save our communities, it's going to be in that first few minutes, in that first hour.

And I applaud all the work that's been done by the DOD; I applaud all the work that's been done by government agencies. The bottom line is if I have to wait 3 or 4 hours, then send me a lot of body bags because that's what's going to be left.

I think that we need to emphasize about the enhancement of existing local forces. I have heard a lot of nice things here today about local responders. They need to be made better, more professional, more capable because if they don't handle it, it's truly going to be a disaster.

Thank you. I'm open for any questions.

Mrs. FOWLER. Thank you.

I want to thank both Chief Eversole and Councilwoman Simank. Both have presented very interesting testimonies, and raised questions that are the reason we are having these hearings. Hopefully, as a result of the hearing today and others we are going to be having, we can come up with some answers. Solutions that won't take 18 months to get the effect. Chief, I agree with you. I think the problem is now, not down the road, and we need to get you those resources.

I just have a couple of questions. This one is for both of you. As you look at all the different Federal agencies involved, and as we've said, right now there doesn't seem to be one that's really coordinating and in charge, designating one that's not getting the resources or authority it needs.

Which, in your opinion, of the different Federal agencies involved in antiterrorism activities is the best suited to organize and to rationalize and pull together all of these Federal preparedness programs? I mean, which one would you think from the dealings you have had, would be the one that we most need to pull this together? Do you want to address that, first, and then I will go to the councilwoman. Either one, who wants to answer first?

Mr. EVERSOLE. Go ahead.

Ms. SIMANK. Thank you, Chief, and by the way I need to know if he needs a job.

This is a tough question. If you—I keep hearing you all ask for which agency, who should it be, tell us which agency. It's a hard question. I will tell you what I have learned in Oklahoma City. Number one, of course rescue and recovery is what we did, but you are dealing with a crime scene. This is a crime scene. Do not forget that. After everything that we went through in Oklahoma City was over, we are still facing trials. McVeigh is going to possibly be brought to the State of Oklahoma and tried by our district attorney in Oklahoma County. If we had not had good law enforcement collecting the evidence, we would not have had a conviction.

The lives that were lost in the bombing of Oklahoma City left over 3,000 relatives and family members. Those people demand justice after a terroristic attack. They want to see that people are caught, that they are held responsible by the laws of this great Nation for their acts and their deeds that they have done by harming innocent people and lives.

So I don't have your answer, but it seems to me that we are going to have to have some sort of combination of rescue and recovery and law enforcement working together. I have not one agency to give you, Madam Chairman.

Mrs. FOWLER. We know there's not an easy answer today, but I'm just questioning, as you've worked through this and you've given us some good points. Thank you.

Chief Eversole.

Mr. EVERSOLE. Again, you asked for a simple answer to a very difficult question. We asked for one-stop shopping, and I was one of the people who was rather blunt with Janet Reno about that. From that came the NDPO. It seems at that time, and it still seems, like a pretty good way for us to go. The design of that was to have Federal agency people there and also State and local people in that agency who can work together to find an answer for very complicated questions.

See, I really don't care who answers my questions, Madam Chairman. I want the right answer. How the Federal Government does that is your business as long as I get the answer.

But I think what's really important are—there are two things that are important. One, whoever you give that to, you better give them a big whip and a chair so they can crack that whip and make everybody jump in line because I find the competition between Federal agencies intolerable and certainly demeaning to the locals.

The second thing is that they need to have a strong oversight, and I know that they are planning on doing a State and local advisory group. I don't want to be told how to do my job. We know how to do it, and we prove it day after day after day after day. My fire department goes out the door 1,120 times a day. I handle 43 hazmat jobs a month in the city of Chicago. We do that all over this country. I happen to be one of the big funny animals, OK; but there are fire departments all over this country that every day provide that measure to their community, police provide it, EMS provides it.

What we need is a strong input. We want to be able to tell the Federal Government this is what we need. Under the Nunn-Lugar, which sounded very good because, yes, we got the $300,000 worth of equipment and we were given some stuff that, one, we have never used; two, I don't know how to use; three, I don't want to use it; and, four, I'd like to give it back. They sent us things that work well for the military. They are not functional in the city. They don't work right. We need input into what we need. We'll tell you what we need and just help us get that.

We find that they design educational programs, and I would give a lot of credit to the people. I went to McClellan at some of the initial meetings, and the military and the civilian people, it was one step from physical violence.

When they started talking about acceptable losses, the civilian people went out of their minds, and they listened and they have changed their programs, but every one of those consortium programs should have an acid test. An acid test is what are they trained there that we can't get locally? Why is the National Fire Academy consistently and routinely overlooked, shot down, budgets cut, wage freezes, hiring freezes, to the point where they are inept, that they can't do their job? We have a system that we know how to turn on training in every State in this country overnight. Get us a program, get going, just fund us enough to get the instructors, and the government just says, no, we want to invent this; no, we want to invent that. You know, if I ask you what time it is, I don't want you to tell me how to build a clock. Just tell me what time it is. I'll know what to do after that. That's where we need the real help.

And apparently this interagency business is so strong and there are so many dollars involved that it just—we become secondary. The problem is a reason for some people's existence. If that is true, then get rid of those people and let me take those dollars and solve the problem that we have in the cities.

Mrs. FOWLER. Thank you. In the interest of time, I'm going to go on to questions by the ranking member. I do want to welcome Congressman Chris Shays. We introduced you in absentia earlier, Chris, as Chairman of your subcommittee, and commented on all the good work you are doing in this area, too. Would you like to make a statement?

Mr. SHAYS. No, thank you.

Mrs. FOWLER. Mr. Traficant.

Mr. TRAFICANT. Very interesting panel, very competent panel, good to hear from you. I have a couple of questions. I'm not going to start with the chief, but I just want to say, secret chemical underwear? I'll get to that. I'll get to that. But I like people who tell it like it is; and I think both of you —from your respective professional disciplines, one is the public official elected, the other is a public servant certainly that performs a service. You've done that and I appreciate it.

I'd like to start with the distinguished councilwoman.

Ms. SIMANK. Thank you.

Mr. TRAFICANT. Can you tell me, and if you cannot directly, can you get back to me —and I ask unanimous consent that the record be left open for this answer —whether or not Oklahoma City officials of any sort, including their police department or any other official, were notified prior to April 19th, 1995, of the following dynamics: That April 19th was an anniversary date of Waco, Texas, and the fiasco there; and in addition that one Richard Wayne Snell was on that date being executed in Arkansas for killing a state trooper, one Richard Wayne Snell, who had been arrested in the 80's for threatening to blow up the Alfred P. Murrah building? Do you know if there was any communication, any advance notice of the unusual circumstances surrounding the date, the 19th, and this information?

Ms. SIMANK. Honorable Congressman, to the best of my knowledge, my fire chief, my police chief, my city manager, my mayor and other distinguished council members that I serve with had no notification of what you have just talked of. I will, however, if you wish, go home and I will research this to see if I could in any way be incorrect, but my answer would be, no, we had no notice of these particular circumstances.

Mr. TRAFICANT. Thank you. As an old sheriff, there's no one better at the crime scene than local officials who understand the turf. The biggest problem that I have seen as a sheriff, the Federal agencies were usually investigating the fire chief and the sheriff and didn't communicate with us. I don't mean that in jest because as crazy as this sounds I think we need somebody in charge here, and I think when it comes to domestic terrorism, it should be the FBI, and I think we have to streamline it.

But in line with that, I have seen incidents where the ATF and FBI weren't talking; they didn't share information. Chief, just basically to the point, you are the head man in the country for your noble profession, and I could see why. You're a noble spokesman. Do you in Chicago, such a major, great American city, have regular communicative, coordinating meetings with Federal entities regarding any of the issues you've heard discussed here today?

Mr. EVERSOLE. Yes, sir. I am very pleased to tell you that we have built what I would consider an excellent liaison between the Federal Bureau of Investigation and ourselves. Through their assistance, we have lined up a—we're lucky enough in the city to have a level three surety lab which can test for us any known chemical or biological agent; and they have special agents assigned or 10 terrorist specialists in the city. We have cross-trained their people in the use of our chemical protective clothing and made that available to them. They have gone out of their way. I am very pleased with that relationship that has been built.

Mr. TRAFICANT. Having answered that now, in all candidness here, is that because of your efforts reaching out—I guess in asking that, from what you see nationally in your position, is this a model that is happening everywhere or is this an abstract thing that's happening in your area because of maybe the more advanced programs, apparatus you put in place?

Mr. EVERSOLE. Sir, in any type of a true relationship it takes two to tango, and they have come our way and we have gone their way, and it is a working relationship. That's very important to us. I do not believe from my relationship with the International Fire Chiefs that that happens that well everywhere.

Obviously, they have an office in Chicago and we cooperate and participate very well together, more so than we ever have in the past. That's also true with our people working with ATF in the arson end of the business. I know that they work very closely there. I don't think that happens in a lot of little towns. Truthfully, I don't think a lot of cities—if you said how do you call the FBI, they'd probably say, well, I don't know, maybe get the yellow pages and we'll try to find a number or they call 411 and ask information.

Mr. TRAFICANT. You familiar with the Federal Protective Service?

Mr. EVERSOLE. Yes, sir, somewhat.

Mr. TRAFICANT. Do you have communication with the Federal Protective Service?

Mr. EVERSOLE. Only on a very limited basis from my role.

Mr. TRAFICANT. Madam Councilwoman, are you familiar with the Federal Protective Service?

Ms. SIMANK. No, sir, I'm not.

Mr. TRAFICANT. That is the agency that is responsible for the protection and security of Federal buildings. I want you to know at the time of the disaster, the tragedy in Oklahoma City, there was one contract guard, not a full-time security, Federal Protective Service agent, responsible for the three Federal buildings in Oklahoma City.

Ms. SIMANK. I heard you say that earlier.

Mr. TRAFICANT. So having now said that —and I'll conclude my questions. I'll have a couple for the record that I ask you to answer, but I want to ask you one last question. What advice do you have to other jurisdictions in America, just briefly, after having seen and witnessed the fiasco there? What advice do you have for us?

Ms. SIMANK. Help us get prepared. Take this money out of Washington, get it out to your cities and towns, to your local responders. Do not —do not fall into their bickering and their turf wars and their divisions and departments that need to be budgeted, and I understand that. I understand that from my personal business that I'm in as well as my city council position with a city of approximately, you know, half a million people.

Do not —do not let them bring that battle into Congress and battle for money when we're needing it out there. Get the training and equipment to us, and as the chief said, let—trust your cities and towns. They, if they're well trained, can handle disasters; and we need the Federal Government—we need you to assist us and assist means assist, not to come in and tell us how to do it.

I will tell you quite literally we were doing well in Oklahoma City that morning and all day long. It was 15 hours later, but when they started flying in and when the regional office came in from the Dallas/Ft. Worth area, I hate to even say the words here I heard around the table, different Federal agencies and department heads cussing. We're trying to save lives out there on the street, and they're having their turf battles.

That needs to stop. We need to let our cities and towns be well trained and have those resources; and again, I'll just reiterate, my fire chief, Marrs, told me before I came up here, he said, Ann, if we could have all of that money out to the local cities and towns, we would do so much better than spreading all that around to the 42 or 45 agencies. Help us with that.

Mr. TRAFICANT. I really appreciate this panel. Thank you for your candidness.

Mr. EVERSOLE. Congressman, if I may make one short comment. Your statement is very interesting about the Federal Protection Services, and as they are looking at the buildings and hardening the defenses around these buildings, when we talk about the left hand not knowing what the right hand is doing, I'm getting complaints from our people that in some cases they have hardened the defenses to the point where we can no longer get our ladder companies close enough to get to the building.

You've protected yourself right out of business with us, sir. So I would suggest ropes or, you know, very thick mattresses —and I say that in jest —but the truth of the matter is, again, the Federal Government is making decisions and not being cognizant of where the real world is, where people have to live and do business; and they need to find a way to get together and talk to each other so we could come up with reasonable answers to protect our facilities, to protect our entities and our citizens and still do it in a way where we can live.

I'm amazed at the ignorance of some people and their narrow thinking in fixing their problem, and it doesn't —it just creates more problems.

Mr. TRAFICANT. Let me say this: The reform we're talking about with the FPS would bring in the local entities and work closely with the Federal Government.

Thank you, Madam Chairwoman.

Mrs. FOWLER. Thank you, Mr. Traficant.

Mr. Terry.

Mr. TERRY. Thank you.

Chief, you mentioned in your testimony, that was probably the liveliest I've heard in my 5 months on the Hill, that there's been four, I presume, Anthrax hoaxes in your city or biological.

Mr. EVERSOLE. It was only one Anthrax, sir. We had several different chemical warfare. We found a guy who was scared to keep robbing banks with guns because he thought—he told the FBI that he thought somebody would want to be Superman and jump up and take the gun and beat the hell out of him and he'd be done.

So he had read in the newspaper all about this. So he decided he would make a bomb, chemical bomb, and he didn't know how to spell sarin. So he made a Serran bomb, but the matter was that people, as he told—as he told us later on, he said he had never felt so much power because when he said, you know, I'll kill y'all, I got this Serran bomb, and it's chemical, it'll kill everybody, and he said people were pushing into the plaster and marble walls trying to get back away from him; and of course, he got the money. He left the bomb, just keep everybody scared and he ran out, and we had to deal with that.

Mr. TERRY. Well, and that's the focus of my question. So I'll interrupt you on that point there.

You're the ones that have to deal with that. So my question, bringing in with the theme of both of your presentations here, to deal with those four incidents, and thank God they were all hoaxes, but if they were real, did you receive the training or your firefighters under your command receive the training because of the programs that the city of Chicago has set up or because of the assistance or potential assistance from the Federal Government in dealing with those type of situations? I want you to answer that in two parts, one as the chief in Chicago and the other one in your capacity as the head of the International Association of Fire Chiefs, which would assume that Omaha and those type of cities are part of it. Is it different between Chicago and Omaha or Oklahoma City?

Mr. EVERSOLE. I can't answer for every city, but I think it probably is. We did a joint procedural program where, in our city, the fire department handles hazardous materials and the police handle bombs; and we're very pleased with that, and then some guy started making chemical bombs and messed up our procedures.

So we had to rewrite our procedures, and we now work very carefully between our bomb crews and our hazmat units, and the FBI fits right in there, and they have a certain thing that they do, and they have furnished us with, I believe, all the information. Anything we've ever asked for they have given us, and I don't have— I have not a complaint about the FBI and the way they have worked with us.

Mr. TERRY. So there is a role for the Federal Government in assisting you as the wheel, making it more efficient, speedier?

Mr. EVERSOLE. That's true, but the bottom line is we're putting our bomb techs and our hazmat techs are doing the job. The FBI comes in and pays for the lab and gets us anything cleared that we need to get cleared; but, you know, the troops, the troops are the local guys, and it's the troops that got to make the decisions and handle the chemicals and handle the bombs and do all those things.

Mr. TERRY. Part of our oversight needs to be to exactly define, what the Federal role is, so we don't get in your way?

Mr. EVERSOLE. That's very well said, sir. Thank you.

Mr. TERRY. Once in a while, a blind squirrel.

Mr. EVERSOLE. If you could just get the rest of the Federal Government to figure that out, you could run for higher office, sir.

Mr. TERRY. I'll conclude with that.

Mrs. FOWLER. Thank you.

Mr. Isakson.

Mr. ISAKSON. Yes. Is it Simank?

Ms. SIMANK. Simank.

Mr. ISAKSON. I'm sorry. First of all, I wanted to tell you, in particular, firemen, law enforcement officials deserve tremendous praise. What happened in Oklahoma City was a tragedy, and all of us wished we'd never watched it; but as I acknowledged earlier, it was your State police that caught McVeigh, and I believe it was your local police that found the axle.

Ms. SIMANK. That's right.

Mr. ISAKSON. That they traced the truck to Missouri. The Federal Government role was in being able to identify a truck through its information bank but nothing in terms of finding it, which brings me to a question. In your National League of Cities' recommendations you commend the NDPO organization. You make reference that they are trying —and I'm putting that in my quotes, not yours—"to coordinate with local officials."

In all I've heard, it would seem to me that with the exception of a Federal crime, which, as you mentioned, terrorism and a crime go hand in hand, they are inextricably tied. With the exception of DOJ's authority, if it was not in the violation of a Federal law, it seems like we should designate the local governing authority, in which such an incident takes place, as the person in charge and everybody else should be an assistance role. Would you agree with that or disagree with that?

Ms. SIMANK. Now, are you speaking of in charge of the criminal investigation?

Mr. ISAKSON. No, I'm speaking of in charge of handling the response to the tragedy.

Ms. SIMANK. Yes, absolutely.

Mr. ISAKSON. It would seem to me, from what I'm hearing, that's the case. Having thought about something I want to ask the chief in just a second, it's a pretty obvious example. If that is the case; and if the NDPOs would institutionalize their local advisory groups in each State, it would seem like then that it would be a very easy chain of command if the authority was vested in the local elected authority where the jurisdiction where the tragedy took place. Their request for assistance would go through the State advisory group to the NDPO, which may be the State emergency management office. The role of the Federal Government would be to provide those services that are needed for that tragedy, rather than come down and try and provide authority over something they probably know little about. They know a lot about the tragedy but not about the locals. Would that seem to you to work pretty well?

Ms. SIMANK. Well, sir, I don't know. You may open up another area when you put authority with the State, getting on out to the local, the city. For you see, Oklahoma City experienced problems. I'll just be quite frank with you. We experienced problems with our State emergency management people.

Mr. ISAKSON. Not to interrupt, I maybe misstated, but my thought process —let's assume for a second the local elected jurispolitical, jurisdictional body is in charge. Their request for assistance goes through the advisory group that y'all commend in the National League of Cities to the NDPO, and the NDPO's job is to deliver, not question whether or not it's needed.

Ms. SIMANK. It sounds like that might work.

Mr. ISAKSON. OK. For the chief, I have to say I know you're aware of this ,but it gives me a chance to acknowledge again, in Atlanta about 2 months ago, we had a firefighter who reacted to, not a terrorist act, but to a crane operator suspended 250 feet in the air over a cotton mill that was a total inferno. He hung from, I think it was, about an 80-foot rope below a helicopter in a 30-mile wind and plucked him out and saved his life. I commend all firefighters and you and what you do for our citizens every day.

You're right, that guy was picked up, loves the big red truck, and I'm sure glad the Federal Government wasn't in charge of that operation. Nonetheless, going back to the question that I asked here, do you think or isn't it true that as far as your constituency in Chicago is concerned, the buck does stop with you in one of these tragedies? They don't want an explanation of how many Federal agencies may be involved or many people it took? Don't they look to you for the immediate response?

Mr. EVERSOLE. Sir, that question is very easily answered, and the simple thing is this. When the State goes back to Springfield, Illinois, and the Federal people go back to wherever Federal people go back to, the fire chief, police chief, the mayor, the emergency managers all stand out on a corner and for the rest of their lives, remember that's where we leave, they are standing there answering the questions, you know, did you do a good job, did he do a good job, how did it work out.

You know, we have nowhere to escape to, and I would just tell you something that was interesting. That was the first time —and he probably hopes it's the last time—that that firefighter ever has to hang from that helicopter the way he did, and those are things that we do every day in the fire service.

Every day we do—I'm amazed sometimes at what my people do, and we handle chemicals. We have the University of Chicago where there's more Nobel prize winners than any facility in the world; and you know, when those doctors are running out of the building because something drastically has gone wrong, they're yelling, it's in there, Mr. Fireman, and we're going, yeah, OK, we'll get that, and we're running in the building where the expert of the world is running out, and we do that regularly, and those are the things that we have to handle.

What we're asking for is to take what we have, a system that really works, a proven system of 350 years of tradition in this country, and just make it slicker and faster, make my wheel go faster and make it last longer. That's what I need.

Mr. ISAKSON. Can I ask one other question?

Mrs. FOWLER. Yes.

Mr. ISAKSON. In your testimony, you did not discount but you certainly questioned whether or not putting more money in RAID teams made sense. In hearing your testimony, RAID, the first R in the word RAID, stands for "rapid" in that acronym; yet you're talking about 3-minute responses or less when you have disasters in Chicago. I know you probably answered this question: Has there been a demonstration yet to your knowledge of the timeliness of these RAID teams to react or the predictable timeliness of those RAID teams to react?

Mr. EVERSOLE. To the best of my knowledge, sir, there has been no demonstration of this. They're telling us they hope 4 hours they can respond anywhere in our area. We think they can get to Chicago faster than 4 hours.

I have some questions that have not been answered yet. I asked you, of 22 members on board, how many will you roll with? Oh, 22. Are you telling me that these people are going to work 7 days a week, 24 hours a day; that they will never have a day off, never be sick, never be on vacation, and there are still questions.

This is a good concept, but there are many things to be worked out yet, and I'm—you know, the proof is in the pudding, sir. Show me the job. Show me you can do the job. If you can't show me, take your garbage home.

Mr. ISAKSON. To that end, last follow-up. To that end, would you spend $7 million on more RAID teams before you'd seen a demonstration?

Mr. EVERSOLE. Would I?

Mr. ISAKSON. Yes.

Mr. EVERSOLE. Sir, I'm from a local government. Do you know what we can do with $7 million? I'll give you a RAID team that will look like Mr. Whiz's galaxy of such special items.

Mr. ISAKSON. Thanks, Chief.

Ms. SIMANK. Sir, if I might make a point on the RAID teams. I know they're promising cities and towns 4 hours. Maybe they can get to Chicago in 4 hours. I don't quite have much trust in all of that. I will tell you that we, our local fire, rescued most of the victims that survived the bombing, and that happened before 4 hours. We did find one or two alive later, but before the first team ever got to our city. By the time the first team got there, we never found another survivor.

Mr. ISAKSON. Well, I agree with you, within 4 hours y'all had had to deal with unbelievable—in Atlanta, when we had the Olympic park bombing, 4 hours would have been too late——

Ms. SIMANK. Four hours is too late.

Mr. ISAKSON. —When you've got people laying and dying all over the city.

Ms. SIMANK. Four hours is too late.

Mrs. FOWLER. Thank you. I'd like now to turn to our distinguished colleague from the Government Reform Committee who really has been working so diligently in this area and paving the way. Mr. Shays, we appreciate your being here, and if you have any questions or statement, whatever, we'd——

Mr. SHAYS. Thank you. I would just first make an observation. When I walked in the door and, Chief, you were talking, I thought you were Congressman Traficant's brother; and my second question, when I got here was I was thinking you were responding to a question, we were in 10 minutes and you hadn't taken a breath.

So I'm really curious to know what the question was, but you were very effective in what you had to say, and I just would, first, before asking you a question, Chief, say that I think you know that our hearts went out to Oklahoma; and we were just very impressed with how the city dealt with this tragedy during and after, and it is nice to have you here.

Ms. SIMANK. Thank you, sir, and I do know that. Everyone in the United States of America I think felt that same way, and we appreciate that.

Mr. SHAYS. We felt like one family; we really did.

Chief, I'm interested to know if you—what kind of protective gear you have, suits, masks, for chemical exposure to protect from chemical exposure, and if you do have this equipment, did the Federal Government help you buy it?

Mr. EVERSOLE. Sir, we're using the Chemrel level A chemical suit which, as far as we're concerned, is as good as you can buy. You know, when you think of all the factors, and we use Scott Air Paks. Scott Air Paks we had. We use about 11,000 cylinders a month just in our regular business, and every firefighter on duty has his own self-contained breathing apparatus.

I don't go out the door if mine's not in my trunk, and I tell people if I have to put that sucker on and use it, you know it really got bad because I'm supposed to be back at the command post, but the government did supply us with 135 chemical protective suits to our specification, and I applaud them for that. The only problem was, under the Nunn-Lugar Act, all of that was supposed to be for training and not for actual response.

Mr. SHAYS. Interesting, interesting.

Mr. EVERSOLE. We're waiting still for the release. There's a significant appropriation that was supposed to be made, and the grant applications were supposed to be out several months ago for this year's funding. Sir, this is June, and we haven't seen them yet, and I need equipment, and now we're 8 months into this year's Federal budget; is that correct?

Mr. SHAYS. Just before turning back to the floor, just this one question: Does this chemical protective gear also protect you from heat and flames as well?

Mr. EVERSOLE. No, sir, absolutely not.

Mr. SHAYS. OK. Thank you. Thank you very much.

Mrs. FOWLER. Thank you, Congressman Shays. It's my understanding I think, Chief, that the NDPO is holding up these grants because they have had so many requests, but like you say, they need to get them on through their process and get it done; and that's another example of the problems we're having with this setup right now of how the organization is floating up and working very speedily.

I just have two last quick questions because I wanted to go to the other members. Just one to you, Chief. Are you more worried about the training and education level of our responders in a large city or in our remote towns? I know you're in a large city, but if you look at Illinois, there're a lot of small towns around there. Are you more concerned about their training and level of response, education, versus those in large cities?

Mr. EVERSOLE. I think that we need to look at something very intelligently, ma'am. When we had the bad incident in Waco, Texas, most people didn't even know where Waco, Texas, was. When they were up in Montana, who knew where that was? The Unabomber was born and raised less than 10 minutes from my home, but he was in a little cabin out in the middle of nowhere.

I'm sure that McVeigh did not build that bomb in Oklahoma City. He built it in some remote place, and it is the feeling of the International Association of Fire Chiefs that every emergency responder in this country should be adequately trained. Not everybody has to be a brain surgeon or a technician here, but everybody has to be trained to an appropriate level.

We have a system in place that determines awareness, operations, technician, incident, command. We have requirements for that. We know where all that goes, and other people just sit there and say, well, we'll hire some more consultants and we'll build— the program is there, ma'am. We can turn it on tomorrow, and it's just ignored by the Federal Government.

Mrs. FOWLER. Thank you. Are there any other questions from members of the subcommittee here? I want to thank both of you. I'm a former city council member in Florida myself so I understand you're on the front line and dealing with these every day. Again, as you know, our prayers were with everyone in Oklahoma City, and we are hoping through our efforts and those of others working on this that we can, one, avoid those types of instances in the future; and two, make sure we do the best we can to avoid what happened in your city when those groups came in 15 hours later, that we don't have repeats of that either.

And, Chief, thank you. You do an outstanding job, as every member of every fire department throughout this country does. We deeply appreciate it, and you have given us some great testimony today. You both have given us some good suggestions and good ideas, and just know we're going to persevere on this and try to get both of you the most help we can.

So thank you very much again for being with us today.

Mr. EVERSOLE. Thank you.

Mrs. FOWLER. We would like to call our next panel up. Thank you.

Mrs. FOWLER. I would like to welcome our next panel. We'll keep moving. I think we have an hour before the next set of votes. We have Ms. Catherine H. Light, director, Office of National Security Affairs for the Federal Emergency Management Agency. We have the Honorable Charles L. Cragin, who is the acting assistant secretary of Defense for Reserve Affairs for the Department of Defense. And Mrs. Barbara Y. Martinez, who is the deputy director of the National Domestic Preparedness Office for the Federal Bureau of Investigation.

I think you were all here earlier, so you know we have a process of swearing in. So if you would all stand and raise your right hand.

[Witnesses sworn.]

Mrs. FOWLER. Thank you. Be seated.

TESTIMONY OF CATHERINE H. LIGHT, DIRECTOR, OFFICE OF NATIONAL SECURITY AFFAIRS, FEDERAL EMERGENCY MANAGEMENT AGENCY; HONORABLE CHARLES L. CRAGIN, ACTING ASSISTANT SECRETARY OF DEFENSE (RESERVE AFFAIRS), U.S. DEPARTMENT OF DEFENSE; BARBARA Y. MARTINEZ, DEPUTY DIRECTOR, NATIONAL DOMESTIC PREPAREDNESS OFFICE, FEDERAL BUREAU OF INVESTIGATION, U.S. DEPARTMENT OF JUSTICE

Mrs. FOWLER. We'll start out with Ms. Light. If you could begin.

Ms. LIGHT. Thank you, Madam Chairwoman and members of the subcommittee. On behalf of Director James Lee Witt, thank you very much for this opportunity to discuss the status of FEMA's terrorism-related activities. As you know, we have provided a written statement; and I'll now summarize some key points from that statement.

First, I'll give an overview of FEMA's responsibilities in terrorism preparedness and response, then briefly describe FEMA's role in consequence management and discuss FEMA's programs and activities to better prepare States and local governments for dealing with terrorist incidents.

While FEMA is the Federal Government's lead agency for consequence management in response to domestic terrorist incidents involving WMD, the FBI is the lead agency for crisis management.

community in coordinating our activities and programs as part of the overall Federal effort.

And we're committed to doing everything we can to better prepare the States and local jurisdictions for dealing with this immense challenge.

Thank you again for the opportunity to address this subcommittee.

Mrs. FOWLER. Thank you. Secretary Cragin.

Mr. CRAGIN. Thank you. Good afternoon, Chairman Fowler, Mr. Traficant, Chairman Shays. My pleasure to be with you this afternoon.

I would like to discuss the activities of the Department of Defense as a participant in the Federal Response Plan, which Ms. Light referred to in supporting State and local officials in response to incidents involving weapons of mass destruction.

Since President Clinton signed presidential decision directive 62 in May of 1998, significant progress has been made in our efforts to better support State and local authorities. PDD 62, also known as the Combating Terrorism Directive, highlighted the growing threat of unconventional attacks against the United States, and it detailed a new and more systematic method of fighting terrorism here at home by bringing a program management approach to our national counterterrorism efforts.

The directive also established, as you are aware, within the National Security Council, the Office of the National Coordinator for Security, Infrastructure Protection and Counterterrorism to oversee these efforts.

Under that oversight aspect, as Ms. Light has mentioned, two Federal agencies have been designated as the lead Federal agencies, the agencies in charge to handle crisis management, the Department of Justice, FBI, and consequence management, the Federal Emergency Management Agency.

Last year with respect to DOD activities, Congress authorized the creation of 10 National Guard Rapid Assessment and Initial Detection teams. Each RAID team will be composed of 22 full-time National Guard personnel.

These RAID teams are designed to be both State and Federal assets that can initially be deployed by governors. And each team will perform three vital tasks: first, they'll deploy rapidly to help local incident commanders identify and assess the nature and extent of a suspected radiological or biological or chemical event.

I might insert parenthetically in the Tokyo subway attack it was over 3 hours before the Tokyo officials were able to ascertain that they were in fact dealing with sarin gas.

Second, the teams will provide technical advice to civilian first responders regarding such matters as NBC symptoms or casualties, the appropriate decontamination procedures and risk management procedures.

As we look at this map that you have placed up there we see a lot of America that doesn't have the sophistication of an Orange County or a New York City or a Chicago when it comes to dealing with these issues. And where can we say that the next event will occur in America.

And third, these teams will facilitate requests for assistance by identifying needed follow-up by military support and coordinating with local, State, and Federal emergency managers for the utilization of these military support elements.

Another very important mission of the RAID teams once they are stood up, and they are still in training and will not be, in fact, ready to deploy until January of 2000, is that they will continuously work with the first-responder communities to develop and improve emergency response plans and procedures and train and exercise with the local incident commanders who will, in fact, be in charge in the event an incident occurs within their jurisdiction.

The personnel for these 10 teams have been hired. They're undergoing very rigorous training. In August, they'll begin training together as units. And, as I say, we expect to have them certified for deployment early in January of—of 2000.

The Department's FY 2000 budget includes funding to increase the number of RAID teams from the 10 currently authorized to 15. However, you may be aware that the Senate Armed Services Committee recently acted to authorize a total of 27 RAID teams, 17 more than the 10 currently authorized and 12 more than the Department of Defense requested.

The House Armed Services Committee is waiting for the Department's request for full-time positions, which was time staggered as a result of the Defense Authorization Act of last year, and we'll be filing that request the first of July when permitted.

So, really, for the year 2000 we really don't know exactly how many RAID teams are going to be authorized throughout the United States. And a decision on stationing of these new RAID teams will be made once Congress has determined the final number to be authorized in the fiscal year 00.

Those teams will be placed throughout the United States according to criteria such as response times, relative to population density, the proximity of other assets to respond to a WMD incident, and the accessibility of airlift transportation to transport RAID team members to incident sites.

And when we established the first 10 teams, one in each FEMA region, one of the criterion was to look to see that there was National Guard organic airlift capability available to provide that airlift. Because, obviously, if you have one in each of the 10 FEMA regions, you're not going to get to a lot of places in that region by getting in your van and heading to the scene immediately.

We're also, as part of Secretary Cohen's program to support local and State officials and to support our civilian Federal agencies who, as I said, are in charge, we're going to be training and equipping 43 chem-bio reconnaissance elements and 127 decontamination elements that are resident in our Army Reserve, our Army National Guard, our Air Force Reserve, and our Air National Guard enabling them to more effectively respond to a WMD event.

In addition, and at the direction of Congress this past year, the Department is working to establish what we were directed to establish, what we call RAID light teams, that John Eversole referred to in his testimony in each of the States and territories where a RAID team has not been located to date.

Last October, Congress also directed that we utilize our existing training technologies to expand the training base available to civilian first responders. We're collaborating with other Federal agencies to adapt and convert WMD preparedness courses so that they may be used throughout our advanced distributed learning process.

Through this shared use concept, these and future courses will be available to both the Federal interagency community and the Nation's 2 million first responders. And as you're also aware—and it's been alluded to a number of times this afternoon—the Department of Defense is administering the Domestic Preparedness Program, the so-called Nunn, Lugar and Domenici Act, which provides WMD preparedness training for 120 of America's largest cities.

The program focuses on providing initial awareness, protection, decontamination, and detection training. It is not the sort of ongoing sustainment training that you heard Ms. Light refer to with respect to FEMA's activities.

To date, we have trained 58 cities involving over 15,700 responders. And as part of this program, we have worked with other Federal agencies to compile a compendium of Federal weapons of mass destruction preparedness courses, the training courses that you mentioned earlier today, so that these courses can be available to State and local agencies.

And a final version was published in July of last year and is available on the Internet. And in the past year, both the Department of Defense and the Department of Justice have conducted a number of forums with first responders. Chief Eversole has been an energetic participant in those forums. And you heard him say that one of the things we need is one-stop shopping.

And frankly, we heard Chief Eversole and his colleagues in the State and local capacities and that was one of the reasons that, one, the Department of Defense agreed with the Department of Justice to transition the Domestic Preparedness Program from the Department of Defense to Justice, so that they would have it in the same location with their equipment grant training programs, hence one-stop shopping.

And then the National Domestic Preparedness Office was established under the auspices of the FBI—and Ms. Martinez will speak to you in detail about that—but as a means to coordinate the provision of services from the Federal Government, not to be the provider of all services, but to ensure that there was a one-stop shop for dealing with State and local officials.

And we're frankly very pleased with the progress that's been made in the past year under PDD 62 and the interagency working groups. And I can tell you, in spite of some of the testimony that you've heard to date, these ladies here and I talk on a regular basis and we talk among our interagency colleagues; and, in fact, we do know each other's phone numbers and use them regularly. Thank you, Madam Chairman.

Mrs. FOWLER. Thank you, Secretary Cragin. Ms. Martinez.

Ms. MARTINEZ. Thank you, Mr. Cragin. Thank you, Madam Chairwoman. And thank you for the opportunity to speak before distinguished members of Congress, my colleagues, regarding the proposed role of the National Domestic Preparedness Office in combating terrorism within the United States.

My intent is to highlight the importance of achieving coordination across the Federal Government of the various individual agency efforts that provide very valuable assistance to State and local communities in preparing them to face a challenge that terrorism presents.

As over 40 Federal agencies would have a role in the response to a terrorist event, so too are many of them in a logical position to provide valuable assistance to prepare our communities before such an event occurs. The mission of the proposed National Domestic Preparedness Office consistent with the recommendation made to the Attorney General by State and local authorities will be to serve as that central coordinating body for Federal programs that provide preparedness assistance to State and local communities regarding terrorist incidents, particularly those involving weapons of mass destruction.

As you know, in the past few years the President of the United States and Congress have taken significant steps to increase our national security and to promote interagency cooperation. Most recently, cooperative efforts against terrorism have been extended to include State and local agencies, professional and private sector associations as well.

For example, in the preparation of the administration's 5-year interagency counterterrorism and technology crime plan, the Attorney General directed the Department of Justice Office of Justice Programs to host a meeting of individuals like Chief Eversole who represent the various emergency response disciplines that would most likely be involved in the response to a terrorist incident.

More than 200 stakeholders representing local and State disciplines of fire services and hazmat personnel, law enforcement and public safety personnel, emergency medical and public health professionals, emergency management and other government officials, as well as various professional associations and organizations, all attended the 2-day sessions.

Collectively, they made recommendations to the Attorney General, James Lee Witt, director of FEMA, Dr. Hamre, the Deputy Secretary of Defense, and other Federal officials on ways to improve Federal assistance for State and local communities. These recommendations have been incorporated in that secret administration's 5-year plan that I mentioned.

The most critical issue identified by stakeholders was the need for a central Federal point of coordination. Due to the size and the complexity of both the problem of terrorism and the Federal Government itself, it was no surprise that the many different avenues through which aid may be acquired by State and local officials and the potential inconsistency of those programs was deemed simply overwhelming.

In essence, the Federal Government, though well intentioned, was not operating in an optimal manner, nor was it effectively serving its constituents with regard to domestic preparedness programs and issues.

State and local emergency response officials made the strong recommendation to the Attorney General for the coordination and integration of all Federal assistance programs that reach State and local authorities for terrorism preparedness.

In heeding these recommendations, the Attorney General consulted with the National Security Council, the Department of Defense, the Federal Emergency Management Agency, Department of Health and Human Services, the FBI, as well as other relevant agencies, regarding the establishment of the NDPO.

So if approved, this NDPO would provide that single coordination point within the Federal Government to better meet the needs of the Nation. It's intended that the NDPO will serve as that much-needed clearing house to provide information to local and State officials who must determine the preparedness strategy for their own communities.

In keeping with the stakeholders request, the NDPO will also provide a forum for the establishment of an agreed upon standards to guide the execution of Federal programs. Federal participants that will serve in a full-time capacity at the NDPO once approved will include FEMA and the Department of Defense as well as the National Guard Bureau, Department of Energy, Department of Health and Human Services, the Environmental Protection Agency, the Office of Justice Programs, and the FBI.

We have also received commitments from other agencies including the U.S. Coast Guard, Department of Veterans Affairs, and the Nuclear Regulatory Commission to provide personnel in the future.

Stakeholders also cited the need for formal representation of State and local officials along with those Federal agencies in the form of an advisor board to guide the development and delivery of more effective Federal programs.

Federal agencies agree that their participation is critical to the whole process of domestic preparedness; therefore, in addition to the advisory board, it is anticipated that when fully staffed, the NDPO will be comprised of approximately one-third State and local experts from various disciplines on a daily basis.

The stakeholders identified six broad-issue areas in need of coordination, planning, training, exercise, equipment research and development, information sharing, and public health and medical services.

If I have a moment, I would just like to speak on the highlights of some of these programs.

Mrs. FOWLER. Really highlight them because I'm afraid we're going to have votes before we get to our questions.

Ms. MARTINEZ. I understand that and appreciate that. In the area of planning, we'd facilitate distribution of guidance that would explain to State and local planners the logistics of how Federal assets would be included into their local response plans. We will continue the compendium that Mr. Cragin spoke of to catalog the existing Federal training programs.

With respect to the information sharing, we would implement a mechanism to facilitate access to personnel outside law enforcement to information that would be important for preparedness and consequence management including Web sites and other links.

With respect to after-actions from exercises, we will maintain a data base that will allow after-actions from exercise as well as actual events, including Oklahoma City, to assist the efforts of other communities.

Under the guidance of public health service, the NDPO would help coordinate the efforts to support the metropolitan medical response systems and the wholesale integration of the public and mental health care community into the response plans.

In summary, then, over the past few years we have seen growing cooperation at all levels, growing interest at all levels of government to address the preparedness needs of the Nation. However, there is a tremendous amount of work that needs to be done.

We clearly recognize that the actions of the first people on the scene will make the difference between life and death. We also believe that the key is to work together to form a partnership among Federal, State, and local communities to better prepare for a coordinated response that saves lives and provides safety for all involved. So in the interest of time here, I thank you very much for your attention.

Mrs. FOWLER. Thank you. We have your whole statement. I have read it. I know the other members will too because I know we want to keep everyone here before we have votes.

I want to thank all three of you because you have all been working in your areas and many people, too, in these areas.

I have some questions that I would like to ask. And I would like to turn your attention to this poster board over here, because Secretary Cragin—this is a map showing where the DOD training has been done, or is scheduled to be done, in those 10 cities. Can you explain why there appear to be dense clusters of training in certain areas and then there are big holes where no training at all is concerned? What I see is the yellow States, that's 11 States where there is absolutely zero training being done by DOD, but then if you look down in southern California, northern California——

Mr. CRAGIN. Florida.

Mrs. FOWLER. —certain other areas, there will be clusters. Could you explain that to us.

Mr. CRAGIN. Well, take a look, for example, Chairman Fowler, at Florida. There are a number of cities in Florida that have training. And the reason why they do is because what the Department of Defense decided after meeting and conferring with the interagency was that they would go to the 120 largest cities based on the census data in 1990, and essentially that meant that any city that had population in excess of 144,000 would, in fact, receive the training. And it was on that basis that the decision was made.

Now, I should say, and as I think the observations were made by your representatives from the GAO, this has been a very evolving training program. And, in fact, Congress in its wisdom directed the Department on an annual basis to review its training activities and then to reassess the program. So now in many of these training cities there are many other agencies and jurisdictions that are involved and participate.

Mrs. FOWLER. Well, as you heard Chief Eversole mention, that he's very worried about the lack of training in rural areas. We look at the incidents that have been occurring over the past years, some of these very areas that are in yellow is where some of this originated or was occurring.

I'm deeply concerned about what lacks here. And, in fact, we have a chart here that shows a concentration of where our extremist groups are the highest. Here we go, right here. If you look at that chart, where they are the highest appears to be in some of the areas where we have no training. It concerns me, as we are looking at this list, Idaho has the highest concentration of extremist groups in the country; yet if you go back to the other map you see what is happening here. Montana has the second highest concentration of extremist groups, with no training being done by DOD.

And you go on down the list, I'm not sure what—you know, just looking at population rather than at risk and what groups are out there doing what, has caused us some concern as we started compiling our information and looking at where the holes were.

So does that trouble you that some of the most concentrated areas for extremist groups in this country are totally missed by the DOD training program?

Mr. CRAGIN. I guess it doesn't trouble me in the sense, Chairman Fowler, that when the Department of Defense had to make the decision as to how to implement this program, that it tried to take the existing resources available and maximize that utility.

As I said, as we have evolved and grown with this program, we have involved many other jurisdictions in the training aspects rather than just one city. But I think it's like every other judgment that we all have to make, you have to balance all of the competing interests. We chose to do that.

The equipment grant program, for example, is a loan program that consisted of a maximum of $300,000 per city for providing training equipment to those cities. Had we tried to expand the base and utilize—get to more of the population or different geographic areas, we would have been diminishing the focus of the program.

Mrs. FOWLER. Of course we heard from Chief Eversole about some of the efficacies of what they got under that program, but I'll leave them to argue that with you.

Mr. CRAGIN. We can all chat about that at a different time. We have a different perspective.

Mrs. FOWLER. You've led me into my next concern, which goes to FEMA because both FEMA and Department of Justice are coordinating and providing training in about 120 jurisdictions also.

And DOD, if you'll look at this next, DOD is currently providing its training program in 89 of these same jurisdictions. So we did a little chart here because now we've got—you know, there's a huge amount of overlap in certain areas of the country where you have got all three of you doing training in areas and then other States where there is none being done or huge areas of States.

You take my State of Florida, all of northwest Florida has nothing going on, period; and we have the capital there and nothing is happening in all of northwest Florida or in southwest Florida. I just look at my State; I look at Mr. Isakson's State. South Georgia has nothing.

So I'm very concerned and just wondering why there wasn't some coordination. We get right back to the same problem again between DOJ and FEMA, in which FEMA is already training; and then DOD comes in with theirs. There appears to be no coordination. FEMA is already doing it here so we'll go fill in the gaps. Instead, we have duplication occurring in a lot of areas.

Could either one of you address to why there wasn't more coordination, with right hand talking to the left or deciding to cover more of the country rather than less?

Ms. LIGHT. First of all, FEMA is involved in both of the programs. Under one act of legislation, we were tasked to work with DOJ on that program and under the other act of legislation we were asked to work with DOD; and we certainly did that. But the grant money that FEMA has goes out to States and locals—and from our perspective the more money that can go out to States and locals the better off we are.

The money actually goes out through the States so that every State in the country, in fact, does receive some money for training. And it goes out through two different arenas, the State emergency management system and the fire training system which Chief Eversole alluded to just a little bit earlier.

Mrs. FOWLER. But have you some duplication occurring here if we could take the same amount of money that you all are spending together that 89 are getting. I quoted the local official earlier, how many different ways can you tell me to bake the chicken.

One area is going to three different programs or four different programs, whereas another area is getting none. And there didn't seem to be any looking at why are we duplicating versus addressing an area that's getting none in the country.

Mr. CRAGIN. I think, Chairman Fowler, that is an issue that is being addressed. You're absolutely right, there has been duplication. You have only talked about a couple of different actions, but HHS is also engaged in providing medical training to emergency responders through their programs at the direction of their authorizing committees.

The office of Justice programs is part of the development of the memorandum of understanding, and in the transition of the domestic preparedness program to them, Justice has agreed that they will review and evaluate these pockets in the United States that are not receiving any training and try to see if they, as part of a broader program, can meet those needs.

Mrs. FOWLER. OK. But then I'm worried we're throwing good money after bad. You have done 58 cities out of your 120. What you're saying is DOD is not going to reevaluate the remaining cities that you haven't done of that 120. Instead of going in some of these where we're duplicating, maybe we should go in some of these States like Idaho and Montana where there's a great need and do that.

You're going to keep traveling down the road of those original 120 and later let DOJ say, oh, well now some other cities need training and we need to put more money in and more. Is that what I'm hearing?

Mr. CRAGIN. That's our intent.

Mrs. FOWLER. There is no evaluation that is going on?

Mr. CRAGIN. We have made what we believe is a commitment to those cities. They have become engaged in the planning process, anticipating that they will receive that training. And we think that the Federal Government has an obligation to live up to that commitment.

Mrs. FOWLER. I just have one other quick line of questioning for Ms. Martinez. All it takes is yes or no answers because this goes, again, to the problems with the NDPO as to whether it's got the resources authority, the capabilities to do what we hope it can do at some point in time. Does your office have the capability to determine where Federal response teams should be located to optimize the use of Federal resources? Today does it have that?

Ms. MARTINEZ. You're speaking on response teams. We're actually in the business of preparedness, so that wouldn't be us.

Mrs. FOWLER. You wouldn't be doing that. OK. So you cannot do that. You can't decide where the resources should go. Two, does your office have the authority to decide where to locate Federal training courses so we are getting the most bang for our buck and getting training more throughout the country?

Ms. MARTINEZ. If I could offer more than a yes or no to that.

Mrs. FOWLER. OK.

Ms. MARTINEZ. I would like to echo what some of—the previous question. What the NDPO is offering right now without expense to the Federal Government is on the basis of what GAO asked us last year to do, and that which appears in the National Defense Authorization Act of this current year, asked the FBI to conduct a threat and risk assessment on the heels of which will be a capability and needs assessment.

We recognize exactly what you're saying, the duplication and the gross gaps that do exist. So we would like, at the present time, to say that we can move out very quickly with this assessment while this other training program is in the works, and using the time and resources of so many people, we can be doing a parallel effort to fill those gaps with the training that is needed by those people in those gaps.

So at the present time, the FBI is tapped to do a threat-and-risk assessment, and we are under way to do that. I would invite GAO—we have spoken with them——

Mrs. FOWLER. Do you have a time frame when that is going to be completed?

Ms. MARTINEZ. We were expected to develop and test the methodology by September of this year.

Mrs. FOWLER. That's just the methodology. So by the time you get your answers, it will be sometime in the middle of next year.

Ms. MARTINEZ. We expect implementation in February of the next year, yes, in the first and second quarters.

Mrs. FOWLER. So my concern is meantime you've got all these areas of the country not getting the training and all this duplication continuing and $10 billion being allocated for this next year, a lot of which could go to areas where there are going to still be holes and not covered until, as Chief Eversole—that's why they get frustrated at the local level, because we study it to death.

My staff didn't take long to put this together. It doesn't take rocket scientists to see where the holes are and where the duplication is. So I have some real concerns there.

Another question: Does your office have the authority to identify and get rid of duplicative training programs?

Ms. MARTINEZ. No, we won't get rid of duplicative.

Mrs. FOWLER. OK. Do you plan to evaluate the need and effectiveness of the new RAID teams?

Ms. MARTINEZ. No, not at all.

Mrs. FOWLER. So I think those answers give us some ideas of what authority the NDPO is going to have and some of the problems we're facing.

I would like now to turn to Mr. Traficant.

Mr. TRAFICANT. I would like a yes or no on this, not to be offensive. Ms. Light, you know what a Stinger missile is?

Ms. LIGHT. No.

Mr. TRAFICANT. Miss Martinez?

Ms. MARTINEZ. I do know what a Stinger missile is.

Mr. TRAFICANT. What is it?

Ms. MARTINEZ. It's a military operative missile, has heat-seeking capabilities for offensive——

Mr. TRAFICANT. How is it deployed and how is it launched?

Ms. MARTINEZ. Normally with a LAWs rocket-type of effect, to my knowledge.

Mr. TRAFICANT. Can you tell us what a Stinger missile is?

Mr. CRAGIN. You're referring to a hand-held type missile? Can be launched off someone's shoulder.

Mr. TRAFICANT. Off someone's shoulder. See, in 1995 it was fertilizer because that's what they could get. Are we so naive to believe—now I am talking about rocket science here—that we're going to continue to have 1950 culprits in a 2000 millennium age?

The reason I'm asking these questions is I've heard all of this discussion of all these agencies. And quite frankly I think people are falling over each other. I have yet to hear prevention of any significant context. I don't know what we're doing.

The fire chief is going to have to go in after there's a disaster. The police chief is going to have to investigate a crime scene. Now, the reason why I'm asking this, because I am really frosted over an issue. We are investigating and inspecting only three out of every 100 trucks on our border. And a report just come out, an intelligence report said they can literally smuggle a nuclear weapon across our border and we may not even detect it.

Now, when I look at defense and I look at our border, I'm not looking at a local fire chief on the border. I'm not looking at a sheriff from some Texas community on our border. That's a national security issue. And I don't see how in God's name we have any kind of an antiterrorist, terrorist prevention program in place at all without a concerted effort that secures our border, ensures that none of these types of devices can come across our border, let alone someone running across with a backpack full of TNT or a Stinger missile.

So my question is—and you don't even have to answer this—for the military, I want to know why they're resisting helping on the border. Because I think the military understands the issue, and I think the military just wants money to do it, more money.

And second of all, I want to know from FEMA and from FBI—because FBI I think should be in charge of this, domestic terrorism business, FEMA should come in to make sure everybody is made whole after its over—I want to know what prevention programs you have. I want them specifically sent out here, what efforts you have on the border, how you interrelate intelligence between the Department of Defense on border activities and customs points and how you interrelate with customs and border patrol on reported incidents of both narcotics and any potential terrorist organizations doing business in those areas.

These terrorist organizations are very simple. They all finance their methodologies with narcotics. From that whole group, Turkey, the Iranians, they're involved in narcotics. And all of these things now are absolutely ripe on our borders.

And not one incident, Madam Chairwoman, in ever discussing terrorism has anybody significantly talked about our borders that are wide open. So I want you to answer that question.

In addition, I have a number of other questions that I would like to be submitted for the record. I would ask for them to be answered and submitted in writing. Thank you.

Mrs. FOWLER. Thank you.

Mr. Isakson.

Mr. ISAKSON. Secretary Cragin, I wanted to—your explanation of the RAID teams was interesting because in what you were describing. It appeared to me in your description that there were more teams to come because of a suspected, but not actual, event than came in after an actual terrorist event. Did I hear that wrong?

Mr. CRAGIN. They could be utilized in both modalities, Mr. Isakson. Perhaps it would be useful to explain how the concept of the RAID teams came about. And we heard from a lot of first responders around the country in smaller communities that said we really don't have the resident expertise to be able to make some of the initial assessments that we have to make in order to utilize our local resources. And Secretary Cohen heard those comments.

We, the Department of Defense, is a supporting agency to civilian authorities. We provide support. We had a very natural vehicle, however, and that was the National Guard which, as you know, is a State asset unless it's Federalized.

And so what was recommended to the Secretary and what he endorsed was to develop this very limited number of individuals who had a great deal of expertise and who were dedicated to this mission on a full-time basis, place them in the National Guard with full-time National Guard personnel working there, and giving the governors the ability to deploy them immediately, rather than have to work through the Federal Stafford Act bureaucracy or any other bureaucracy to get a declaration.

So we developed the first 10 teams. As I mentioned, they're still in the training process. We haven't even deployed them yet. We were still learning as we moved. We did an analysis of the country after stationing one in each of the 10 FEMA regions and found that we had basic pockets where we knew there was no way, given the criterion, that we could deploy them effectively to certain pockets.

So we came in as part of the President's budget and requested authority for an additional fielding of five teams which would be, obviously, 1–1/2 behind those teams in training. But they can go to the scene at the request of State and local officials before an event occurs, or they can arrive after an event occurs but also have substantial reach-back capacity.

You know, we have talked a lot about chemical today. We haven't said word one about biological, which is an entirely different situation. And I appreciated your threat-risk assessment on the continuum.

These folks are going to be trained to come in and make assessments as to what is the biologic—you know, was it Anthrax and boctulinum at B'nai Brith in downtown Washington? We didn't really know at that instant.

We need to be able to have this sort of expertise. We can't have it in every town in America. We just can't afford that resource. So we have to try to maximize what we have and then make it as mobile and responsive as we can.

Mr. ISAKSON. The teams are 22 men each; is that correct?

Mr. CRAGIN. Men and women, yes, sir.

Mr. ISAKSON. If you go to 15, you would be within 4 hours of any domestic location; is that correct?

Mr. CRAGIN. No, we would not. Twenty-seven teams would give about 95 percent coverage of the population of the United States including the non-CONUS States.

Mr. ISAKSON. Did I hear right, 4 hours? Is that the goal?

Mr. CRAGIN. That's the goal.

Mr. ISAKSON. And the 22-member men and women teams are full-time Guard employees.

Mr. CRAGIN. They are.

Mr. ISAKSON. Both civilian and——

Mr. CRAGIN. No, they're all military.

Mr. ISAKSON. All military.

Mr. CRAGIN. Yes, sir.

Mr. ISAKSON. One other comment. I notice Marietta, Georgia, is one of the original 10. That is Dobbins, I take it.

Mr. CRAGIN. That is correct.

Mr. ISAKSON. I would like if it's permissible to go out and visit, since I'm on this subcommittee, the benefit of just seeing where they are on that training and how that's going, if that's possible.

Mr. CRAGIN. I would be happy to make that arrangement for you.

Mr. ISAKSON. As I say, they're still in the training program individually. We're taking advantage of a lot of the interagency training from FEMA and the fire academy and other places. But they'll start collective training in August. And I'll work with your staff to make sure that we make that available to you.

Mr. ISAKSON. One last question, Madam Chairperson. I think I heard you say—correct me if I'm wrong—and I'm not trying to set you up, but the fire chief was animated and articulate in his——

Mr. CRAGIN. He is that.

Mr. ISAKSON. —belief that locals can do it better. I happen to share in that belief, not as a criticism of the Federal Government, but it's always easier for the guy closest to home because they have a lot of local knowledge.

However, it's also as he observed, it's resources of the Federal Government that make some of the advancements we have possible in many things, whether it's medicine, defense, what's going on in Yugoslavia, everything else.

I thought I heard you say that—or at least in what you said that the agency—the RAID teams were at the disposal of the governor to serve in the State where an expected—or where a tragedy may be about to happen or one had happened. Is that correct?

Mr. CRAGIN. That's correct.

Mr. ISAKSON. Does that mean that they, those 22 men and women, are under the authority of the governor of that State to serve that local community?

Mr. CRAGIN. It means——

Mr. ISAKSON. When they're on the site.

Mr. CRAGIN. That's right. They are in a support role to the local incident commander who is the chief of police or the fire chief. And that's all they are there for. And in fact, Mr. Isakson, if we deploy—in the Atlanta Olympics, I happened to be a naval reserve officer who was part of the weapons of mass destruction task force that was down at Fort Gillem.

If we had stood up that entire task force, it would have been there to support the lead Federal agency which would be there to support the local incident commander.

In the final analysis we understand where the responsibility is in this Republic. It starts at the local community, it works its way up to the state level, and then at some point the Governor of the State says the resources of this State have been exhausted and we need assistance from the Federal system.

But also with the RAID teams, just keep in mind there's only 10 of them, and we have 48 contiguous States. We looked at mutual aid compacts to ensure that we could move a RAID team, from, say, Georgia into North Carolina if called upon through a mutual aid contract.

Mr. ISAKSON. Well, I appreciate the answers and I will visit. I would say that what Councilwoman—is it Simank—said, is pretty typical oftentimes of when Federal agencies do come into a local area. All of a sudden they get into arguments or they get into turf protection or it's not as cooperative as it ought to be.

And in a tragedy like an anticipated, but not yet happened, biological or nuclear or contemporary weapon event or the aftermath, I do think those points are well taken. You've got to be able to respond to the local people who know the community and the needs best on the ground. I think they were right.

Mr. CRAGIN. That's also what these RAID teams are going to be doing. They're going to be full-time people who will be training with the local responders, who will be exercising with the local responders, who will learn the incident command system from the get go and, in fact, live in the communities as National Guard personnel.

Mr. ISAKSON. Thank you very much.

Mrs. FOWLER. I just have two final questions: one, you were just referring to the mutual aid compacts. My understanding—I think there are only so far 33 States have those. In looking at this list of where your first 10 teams are, two of them are located in States that haven't signed them.

So my understanding is that the RAID team in Massachusetts, if there was an incident in Connecticut, couldn't go over to Connecticut to help. In California, you've got one in California that's not a part of any mutual aid pact.

Crisis management focuses mainly on law enforcement activities related to the causes of a threat or actual incident.

Consequence management focuses on the effects of an incident. It includes measures to protect public health and safety, support essential government services, and provide disaster and emergency assistance to an affected area.

FEMA uses the Federal Response Plan, or the FRP, as the vehicle to coordinate Federal consequence management activities. Over the years, the FRP has been used in numerous presidentially declared disasters and emergencies. The plan brings together 27 departments and agencies to organize Federal disaster response and recovery efforts in support of State and local requirements.

Most importantly, the FRP provides a known and flexible framework under which local, State, and Federal officials can orchestrate their response and make the most effective use of the available resources.

In implementing its domestic terrorism preparedness activities, FEMA strives to ensure several things: first of all that State and local responders and emergency management personnel are the focus of the Federal programs; that needs of the balance of the Nation, not just the largest cities and metropolitan areas, are addressed; that initial training is reenforced and sustained with refresher information and updated instruction; and that existing plans, capabilities, and systems are utilized as the foundation for addressing the unique requirements of WMD.

FEMA Director Witt has been working very closely with the Attorney General to better coordinate Federal interagency efforts, including support for the National Domestic Preparedness Office. In addition to supporting that office, FEMA will continue its lead agency responsibilities for consequence management.

In FY '99, FEMA is making available over $12 million in grants to States and local jurisdictions. This includes a little over 8 million for use by State emergency management agencies for planning, training, and exercises and 4 million for use by the State fire training centers to support the delivery of terrorism training programs.

With respect to planning, FEMA applies experience gained in responding to natural disasters to the development of terrorism and consequence management plans and procedures. In 1997 we published the Terrorism Incident Annex to the FRP and continue to work with local, State, and Federal interagency community to refine our response.

In the area of training, FEMA has developed and delivered a number of terrorism-related courses utilizing existing networks and facilities. In particular, we rely on our National Emergency Training Center, which includes the National Fire Academy, and the Emergency Management Institute as well as State fire and emergency management training systems to deliver terrorism related training to State and local responders.

Regarding exercises, FEMA is working closely with the National Domestic Preparedness Office and the FBI and the States to ensure the development of a comprehensive exercise program. With respect to equipment, we helped develop the standardized equipment list. FEMA is very committed to work with the Federal interagency

So I'm a little concerned that we are putting them in areas that haven't signed mutual aid pacts. Particularly, when you get up in the northeast where you already got a lot of blank spots anyway, and then we have a RAID team in a State that's not part of the mutual aid compact that can't go help Maine or Connecticut or whatever when you're so close.

Mr. CRAGIN. We have an answer for that.

Mrs. FOWLER. Good. I'm glad.

Mr. CRAGIN. First answer is just because a State has not signed a mutual aid compact does not mean that a governor can't authorize a RAID team to go across State lines. They can do that. However, in the event—and I mean, you can consider the scenario and I'm from Maine so I'll use Maine and Massachusetts. We have a problem in Portland, Maine, and Governor King calls up Governor Cellucci and says, send me the RAID team. And Governor Cellucci says, well, wait a minute. If you have a problem in Portland, maybe we're going to have a problem in Boston. Well, at that point that team gets Federalized because that is the beauty of having National Guard personnel who wear a State hat and Federal hat. And the Federal Government sends them to Portland, Maine.

Mrs. FOWLER. So you can override whether they have compacts or not.

Mr. CRAGIN. That's right.

Mrs. FOWLER. Interesting. Another question for FEMA. On paper when you look at all this, FEMA is the lead Federal agency, as you said, for consequence management in the event of a terrorist attack. But as we go through all of this, it appears that FEMA has been providing very little leadership.

It just seems FEMA is one of a number of agencies out there that have training programs and expertise but doesn't seem to exercise much leadership authority nor getting a very significant amount of resources through this.

Could you tell me why this is the case and what would it take for FEMA to assert the leadership it has on paper and make it a reality. Because if on paper you're supposed to be doing this but in reality it's not happening, what is it going to take to make it happen?

Ms. LIGHT. From a consequence management perspective, I believe that FEMA is exercising leadership with respect to that. As you know, the Federal Response Plan is the plan that we use to respond to all type of disasters. So from that perspective we get training opportunities, we get exercising opportunities, we developed a terrorism incident annex to the Federal response plan to focus specifically on terrorism and have provided that to States and to locals. And we are working very closely with the FBI to dovetail the lead agency responsibilities for crisis management and for consequence management.

And in the area of training, it's true we are in some capacities definitely supporting DOD in the Nunn-Lugar program and supporting DOJ in the DOJ FEMA program. But I think we are also taking the initiative to make good use of the money that we do have available. Of the money that we get, two-thirds of it actually goes out to the States.

And we are taking the opportunity to use that money to provide it to those States and localities who are not, in essence, receiving the training through the other programs.

Mrs. FOWLER. Well, I would just like to recommend that each of you read Councilwoman Simank's testimony if you didn't hear it. Because according to it, as well as her statement that she gave here, the problem got worse due to the Federal response plan; that they were doing much better before the Federal teams came in than after.

I'm just saying read it, and, you know, I'm concerned. But I think we need to move on from there. I do want to thank all of you. Because you have really given very good testimony. I spent a lot of time on it. This is going to be an ongoing topic that we're going to be working on.

I think we know that from Oklahoma City, that we heard today, to New York City that we're all familiar with; we do have to prepare to confront domestic terrorisms.

What we have heard today from several of our witnesses is that the Federal Government's current complex organizational structure may be contributing to confusion and resulting in some unnecessary duplication and waste.

The executive branch has taken a number of steps to improve the overall development of this growing array of agencies' and offices' efforts to combat terrorism. But I'm not satisfied that the Nation has a comprehensive plan or clear priorities, and I think that's what we're hearing today from our other witnesses.

So for our support, for Federal support, to be effective—and I don't mind however many dollars we need to spend because this is a critical problem facing our country—but we have got to have a comprehensive national strategy that defines a concrete end-state and is based on some valid assessments of threat and risk of terrorist attacks and what our current capabilities are.

Because without it, we might not be targeting these resources properly. When one of the larger questions you kept hearing from us today is who is in charge, it doesn't appear that any one agency, even though on paper they're saying it's going to be yours, but does anyone have the authority? That's what it all comes down to in this town is who has the authority to control or eliminate waste and unnecessary duplication.

It's important that this cross-cutting delineation of authority be developed—and I think these charts certainly showed it—in order to direct some coordinated efforts. Each time we talk, when one of these tragic events occurs, we all assume our firemen, police, medics are ready. But we have heard today and we have seen this in large parts of our country they may not be ready because they're not getting the training and the resources that they need.

And we've got to make sure that every Federal dollar counts so that the protection that we're taking for granted is actually there for everybody in this country. Since our subcommittee has oversight and legislative authority over Federal emergency management issues, and we're going to be continuing to take a closer look at these questions in the coming months. We look forward to continuing to work with you as we try to make sure that we do keep this country hopefully ahead of it and make sure that in the case—

I agree with my ranking member—we want to prevent these attacks. To make sure that every effort is being done to prevent them. If they happen to occur, make sure that every part of this country has people in the State and local level who are prepared. At the Federal level, when our people go in there to help, they help and not to hinder.

Thank you very much for what you have done. We can look forward to continuing to work with you. I thank Mr. Isakson who has stuck through this whole hearing today. I appreciate the work of the staff. They worked very hard in putting all this information together. And we will continue to be working on it throughout the summer. Thank you. The hearing is adjourned.

[Whereupon, at 5:12 p.m., the subcommittee was adjourned.]

STATEMENT OF

HONORABLE CHARLES L. CRAGIN

ACTING ASSISTANT SECRETARY OF DEFENSE
(RESERVE AFFAIRS)

BEFORE THE HOUSE SUBCOMMITTEE ON
OVERSIGHT, INVESTIGATIONS, AND EMERGENCY MANAGEMENT

PREPAREDNESS FOR TERRORISM RESPONSE

JUNE 9, 1999

Good morning, Madam Chairman and distinguished members of this Subcommittee. Let me begin by thanking you for inviting me to discuss the role of the Department of Defense in supporting the nation's domestic emergency preparedness to respond to incidents involving weapons of mass destruction (WMD).

In the wake of the bombings at the Murrah Federal Building in Oklahoma City and the World Trade Center in New York, it became readily apparent that we as a nation were less than well prepared to respond to terrorist incidents involving WMD. As a result, President Clinton has undertaken significant efforts to galvanize federal agencies and prompt them to work more effectively, both together at the interagency level and in support of first responders, to provide our nation with an enhanced, flexible and integrated response capability.

As a nation, we are also facing the fact that the front lines in the war against terrorism are no longer only overseas—they are also right here at home. As Secretary Cohen has said, we must face the fact that "the next terrorist attack will come to U.S. soil in a bottle or a briefcase." I believe the heightened security measures undertaken during the recent NATO conference here in Washington provides ample evidence of our concern for terrorist activities right here at home. We are determined to ensure that we are prepared for a deadly chemical or biological attack against our country. A comprehensive and coordinated government-wide interagency effort is now underway.

Under the direction of President Clinton, and in partnership with Congress, plans, policies and laws are being developed or revised to help us prepare better for the day when terrorists or rogue nations attack with unconventional means. President Clinton believes we must do more to protect our civilian population from the scourge of chemical and biological weapons, and that we must prepare better to respond to attacks against our Homeland. In May of 1998, in his commencement address at the Naval Academy, the President announced that the government would do more to protect our civilian population from these threats.

Specifically, the President has signed Presidential Decision Directive 62 (PDD 62)—the Combating Terrorism directive—which highlights the growing threat of unconventional attacks against the United States. In essence, PDD-62 helps bring a program management approach to our national counter-terrorism efforts; it details a new and more systematic method of working together to fight terrorism here at home.

PDD-62 established the Office of the National Coordinator for Security, Infrastructure Protection and Counter-Terrorism to oversee national counter-terrorism efforts. This National Security Council (NSC)-directed framework is bringing a new impetus and a new urgency to our efforts to support state and local authorities.

The NSC-chaired Weapons of Mass Destruction Preparedness (WMDP) senior management group coordinates interagency WMDP policy issues and oversees the activities of seven subgroups. These subgroups are engaged in coordinating policies involving federal assistance to state and local authorities, research and development, prevention of WMD from entering the US, security of US WMD facilities and materials, contingency planning and exercises, legislative and legal issues, and intelligence. Each subgroup membership is comprised of the appropriate federal agency/department principals and/or their senior level representatives who can accept or deliver tasks for action.

The Department of Defense (DoD) is an active participant in all of these subgroups, which at its core, operates on the assumption that disaster response is primarily a mission for state and local authorities. As Deputy Secretary of Defense Hamre emphasized during his testimony before the Senate Armed Service Committee on March 9, the role of the Department of Defense is to support other federal, state and local civilian agencies and officials.

Within the DoD, Dr. Hamre issued an internal management plan for implementing its responsibilities as outlined in PDD-62 and to better coordinate DoD-wide WMDP activities. This management plan identified DoD senior management committees and subject matter subgroups that mirror the PDD-62 committee and subgroup structure established by the National Security Council (NSC).

I am responsible, along with a representative of the Secretary of the Army, for coordinating the Department's WMDP efforts involving assistance to state and local authorities, and for representing those activities at the National Security Council's interagency Assistance to State and Local Authorities Subgroup. I also help to coordinate WMDP activities Department-wide and participate on the NSC's WMDP senior management committee.

PDD-62 and the implementing guidance clearly provided the interagency with a more rigorous management structure for coordinating and promulgating national domestic preparedness programs and policies. As always, however, our efforts are designed to support—not supplant—the efforts of state and local agencies and first responders.

The Role of DoD in consequence management

I'd like to take a minute to discuss the Federal Response Plan and the Department of Defense's role in supporting it. The Federal Response Plan (FRP), published by the Federal Emergency Management Agency, is the federal government's manual for coordinating and employing its emergency management response assets to help support state and local responders in managing the consequences of natural and manmade disasters.

The effects of a truly devastating WMD incident likely will overwhelm even the most well-trained and equipped local and state emergency management assets. When that occurs, the affected state governor(s) will then request federal assistance from the President. This usually results in a Presidential declaration of the incident as a national disaster, which will serve as a triggering mechanism for marshalling needed federal support and assistance to state and local authorities.

Under the FRP, the Attorney General of the U.S. (for crisis management) and the Director of the Federal Emergency Management Agency (FEMA) (for consequence management) coordinate the specific federal response assets needed to support a national disaster. The Attorney General, via the FBI, will coordinate federal assets to prevent and deter use of a WMD on United States soil and to apprehend and prosecute terrorists who have perpetrated a WMD attack within the United States. FEMA will coordinate federal assets needed to support local responders in mitigating the consequences of a WMD attack. Federal agencies and the American Red Cross provide support in 12 different emergency response areas—as requested by the lead agencies designated in the FRP. The FRP has been used for over seven years to help coordinate federal support for natural and man-made disasters. It has only recently been updated to include guidance for coordination of federal assets in response to terrorist attacks.

It is important to note that the Department of Defense plays an active, but subordinate supporting role in virtually all disaster response aspects of the FRP. The only Emergency Support Function (ESF) for which DoD has primary responsibility is public works assistance, under the Army Corps of Engineers. It is that subordinate role that influenced the Deputy Secretary of Defense to initiate efforts to transfer lead responsibility for the WMD Domestic Preparedness Program to the federal agency responsible for WMD terrorist matters: the Department of Justice.

The world of domestic preparedness and response is highly dynamic. No single agency acting alone can address the problem in its entirety. As a result, we are in the process of deepening our interagency ties and developing a coordinated approach. We at the Department of Defense realize that this approach is necessary if we are to avoid confusion, both within the federal government and in terms of our ability to communicate effectively with the first responder community. We are working hard to understand the concerns of state and local authorities regarding the federal role in the process. In many respects we share the same concerns, especially regarding the need for a lead federal agency for WMD and the need for the federal government to speak with one voice on this vital issue.

The Department, along with its federal interagency partners are working hard to ensure that we address problems through a coordinated approach. In September 1998, key representatives from the Department of Justice, the Federal Bureau of Investigation, the Federal Emergency Management Agency, the National Security Council Director (Mr. Clarke) and the Department of Defense met to discuss how these agencies could best work together to combat domestic terrorism. It was agreed in principle that the Department of Justice (DoJ) should assume leadership for implementing the nation's domestic preparedness program. This agreement would have the added benefit of placing responsibility for federally supported WMD training and equipping in one location, as part of the "one-stop" shop consistently requested by first responders. DoD has worked in concert with the DOJ to develop a formal memorandum of understanding that will transfer most facets of DoD's Domestic Preparedness program to DoJ.

Within that framework, the Attorney General has proposed the establishment of the National Domestic Preparedness Office (NDPO), which is up and running at FBI headquarters and is even now furthering the integration of our national response efforts. In fact, just last month, the NDPO conducted a three-day training session in Washington, DC for the FBI WMD field coordinators and provided them information on interagency assets and capabilities.

The Domestic Preparedness Program

The Defense Against Weapons of Mass Destruction (WMD) Act of 1996 (Public Law 104-201) authorized Federal agencies to provide resources, training and technical assistance to state and local emergency management personnel who would respond to a WMD terrorist incident. The Act, sponsored by Senators Nunn, Lugar and Domenici, mandated that the United States enhance its capability to respond to domestic terrorist incidents involving nuclear, biological, chemical and radiological weapons. The legislation designated DoD as the interagency lead to carry out a program to provide civilian personnel from federal, state and local agencies with training and expert advice regarding emergency responses to a use or threatened use of WMD or related materials. This interagency effort resulted in the establishment of the "train the trainer" program we call the Domestic Preparedness Program (DPP). In the planning stages of this program, it was agreed that training priority would be given to the largest population centers of the U.S. This translated into a program plan to provide initial training and preparedness assistance for Domestic WMD response for the 120 largest (according to census data) cities in the U.S.

My office provides policy guidance and oversight of the city training/exercises, equipment loans, and expert assistance program aspects of the Domestic Preparedness Program, while the Assistant Secretary of Defense (Special Operations/Low Intensity Conflict) provides oversight for the annual Federal-State-Local exercise mandated by law for the program. The Secretary of Defense designated the Secretary of the Army as the Executive Agent for implementing the program. The Director of Military Support (DOMS) is the staff action agent and the Commander of the Soldier and Biological Chemical Command (SBCCOM) is the Program Director for the Domestic Preparedness Program. In FY97-FY98, the Department spent $79M in support of this domestic preparedness program, in FY99, the Department will spend $50M, and in FY00 we plan to spend $31.4M.

The interagency continues to support our execution of this program. Specifically, they participated with us in the development of our approach for executing this program, which includes initial visits to selected cities, a week of "Train the Trainer" training for local first responder trainers, including hazardous material (HAZMAT), firefighters, law enforcement, and emergency medical service personnel. Tabletop and functional "hands-on" exercises using chemical and biological scenarios further reinforce this training.

A request for training equipment, valued at up to $300,000, is prepared and submitted by each city to meet their specific WMD response training needs. The cities can choose from among three types of equipment needed for training: personnel protection equipment, agent detection equipment, and personnel/equipment decontamination equipment. Course materials and training aids for trainer use are also included as part of this equipment package. City equipment requests are reviewed and approved by the Department. The equipment is then acquired and indefinitely loaned to each city for their subsequent training use. To date, 58 cities have participated in the training and more than 15,700 first responder trainers have been trained.

Additionally, an annual federal, state, and local exercise is held to improve the integration of federal, state, and local response assets during a WMD response. In FY97, the annual exercise was held in conjunction with the "Summit of the Eight" Conference in Denver, CO (May 97). The FY98 exercise was held in September 1998 in Philadelphia, PA. The FY99 exercise, involving a biological scenario, is scheduled for September 1999 in New York City, NY.

Other component elements of the DPP include the Improved Response Program (IRP) and the Expert Assistance Program. The Improved Response Program involves the performance of technical investigations and exercises geared to improve first responder survivability and response to WMD incidents. The IRP has provided practical real world solutions to problems in WMD tactics, procedures and equipment discovered through the DPP training. Testing to validate equipment protection and detection claims is being conducted as part of the Expert Assistance Program. The Expert Assistance Program also provides support for a national Hotline for emergencies, a Helpline for assistance, and web pages that provide technical information needed by first responders.

Development of the joint plan for transitioning overall responsibility for the DPP to DoJ has gone extremely well. We expect the Memorandum of Understanding guiding this transfer to be finalized later this month. DoD will retain responsibility for the city training and equipping program until end of fiscal year 2000, at which time DoJ will honor the commitment to train the remainder of the originally designated 120 cities. During the fiscal year 2000 transition period, DoJ will coordinate with DoD in the city training-planning phases and will begin to provide grant funding for training equipment. The transition will occur in stages to accommodate existing budgets and program plans.

DoD's focus beginning in fiscal year 2001 will be to continue to enhance the readiness of its WMD response units and installation responders. DoJ will focus on the response at the local and state levels. As a result, both agencies will contribute funding to benefit from the lessons learned from the improved response program of the DPP beginning in fiscal year 2001. Joint planning will be conducted through the Multi-Agency Task Force to coordinate both the improvements of state and local response capabilities and DoD's efforts to enhance its response elements.

Beginning in fiscal year 2001, DoJ will assume funding and programmatic responsibility for the Hotline, Helpline and Internet web site, but DoD will retain funding and programmatic responsibility for the chemical-biological database and the equipment testing program, as these program elements are integral to satisfying independent DoD needs. DoJ will coordinate with DoD in joint planning efforts so that the state and local responder communities will continue to benefit from the expert assistance functions. DoD will enhance its domestic chem/bio response capabilities through the Chemical-Biological Rapid Response Team (CB-RRT) by continuing to train, exercise, and maintain this team.

From joint publications to field manuals, from schools to staff colleges, we are working to embed WMD preparedness procedures and training into the way we do business. This effort is particularly pronounced at those schools that produce qualified personnel to perform WMD functions. The training of the Rapid Assessment and Initial Detection (RAID) teams and other elements will mirror our efforts to work across both Service and interagency lines to develop mutually supportive programs. We are working to ensure that the WMD responders, people at the local, state and federal levels, are prepared to work together to meet the needs of the people affected by a WMD attack. Extensive training will include teaching and course work provided by the Army Chemical School, the Defense Nuclear Weapons School, the Army Medical Department, the Environmental Protection Agency, the National Fire Academy, the US Army Medical Research Institute for Infectious Diseases, FEMA, and the Department of Justice's Center for Domestic Preparedness.

The Role of the National Guard and Reserve in Domestic Emergency Preparedness

One effective means of channeling federal support to first responders will come through the National Guard and Reserve. For example, over half our total military medical capability is resident in the Reserve components. In the event of a WMD event, casualties may be enormous—and we will need to call on Reserve component medical expertise and equipment. The Reserve components, predominantly the Army Reserve, also have more than sixty percent of our military chemical-biological detection and decontamination assets. They will be essential providers of support to state and local authorities in the event of a WMD incident.

The Defense Against Weapons of Mass Destruction Act of 1996 required the Secretary of Defense to "...develop and maintain at least one domestic terrorism rapid response team composed of members of the Armed forces and employees of the Department of Defense who are capable of aiding Federal, State, and local officials in the detection, neutralization, containment, dismantlement, and disposal of weapons of mass destruction containing chemical, biological or related materials."

Consequently, in addition to establishing a single Chemical-Biological Rapid Response Team (CB-RRT), last year President Clinton also announced plans to establish ten Rapid Assessment and Initial Detection (RAID) teams in each of the ten FEMA regions. These RAID teams, comprised of full-time National Guard personnel, are designed to be assets of the Governors as they perform three vital tasks. First, they will deploy rapidly to assess suspected radiological, biological or chemical events—in support of the local incident commander. Second, they will advise civilian first responders regarding appropriate actions. And third, they will facilitate requests for assistance. These ten teams are completing extensive technical training and will begin unit-level training this August. We anticipate that they will be fully mission-capable by January 2000.

Funding to support five additional RAID teams was included in DoD's FY 2000 budget request. Congress must approve additional full-time National Guard positions for these teams. The current Senate Armed Services Committee (SASC) mark-up includes authorization to support a total of 27 RAID teams, 17 more than the 10 currently authorized, and 12 more than requested. The House Armed Services Committee (HASC) is awaiting the Department's request to authorize 110 full-time positions for the new five RAID teams, as required by law. Accordingly, the number of authorized RAID teams for FY 00 awaits conference action.

A decision on the stationing of these new RAID teams will be made once Congress determines the final number to be authorized in FY 00. The new RAID teams will be placed so as to maximize the range of effectiveness for all RAID teams. Response times relative to population density, the availability of other assets suitable to respond to a WMD event, and the accessibility of airlift to transport RAID team members to event sites are factors which will be used to evaluate the appropriate stationing of the RAID teams authorized.

The costs of establishing and maintaining a RAID team are not definitively known, as we are still in the process of training and equipping the initial 10 teams. The FY 00 budget requests $14.66M to establish 5 new RAID teams for the first of a two-year start up period. In an unconstrained environment, a one-year cost of establishing a RAID team is estimated to be approximately $7.2M, and the follow-on sustainment cost is estimated at about $4M per year.

In addition, and at the direction of Congress, the Department is working to establish 44 military support detachments, which we refer to as RAID (Light) teams. These teams are being established as part of our overall effort to develop a nation-wide response capability that has strong roots in the local and state first-responder community. They will be established using traditional National Guardsmen and will be built on the RAID model but tailored to the specific needs of each of the States and territories where a RAID team was not placed. The RAID (Light) teams will be structured and trained to provide a modest planning and assessment capability in every state and territory.

To better harness these inherent capabilities and make our national plans for supporting an effective WMD response, each of the Reserve components is being called upon to play an expanded role. The Department of Defense in FY99 and FY00, will train and equip 43 Nuclear, Biological, and Chemical reconnaissance elements and 127 decontamination elements in the Army Reserve, Air Force Reserve, Army National Guard and Air National Guard, enabling them to more effectively respond to a WMD attack.

Use of Training Technologies

Last October, Congress also directed that the department take advantage of existing training technologies to expand the training base available to civilian first responders. The Department is currently converting established courses so that they may be used through the Advanced Distributed Learning (ADL) process. The Department is also collaborating with other federal agencies to adapt their appropriate preparedness courses for this training venue. Through this shared use concept, these and future courses will be available to both the federal interagency community and the nation's two million first responders.

Once the conversion process is completed, initial pilot training sessions will be undertaken through the National Guard's Distributed Training Technology Program infrastructure, as well as the Reserve Components Assessment System (RCAS), Warrior and civilian emergency response/training networks. These pilot training sessions are currently scheduled for the third and fourth quarters of fiscal year 2000. The ADL approach compliments the conventional training process, provides increased learning effectiveness and greatly increases accessibility for students, while reducing training time and costs.

As part of the Domestic Preparedness Program, the Department of Defense worked with other Federal agencies to compile a compendium listing of federal Weapons of Mass Destruction preparedness courses available to state and local agencies. Courses offered by the Department of Defense, Department of Energy, Department of Health and Human Services, Federal Bureau of Investigation, Federal Emergency Management Agency, Environmental Protection Agency, Department of Justice, Department of Transportation, Emergency Management Institute and the National Fire Academy are listed. A final version was published in July 1998 and is available for viewing on the Soldier Biological and Chemical Command's Domestic Preparedness web page: ***www.nbc-prepare.org.***

Future updates to the compendium will be accomplished by the FBI's NDPO in keeping with the "one stop shopping idea" advocated by the Attorney General and the nation's emergency response community. The NDPO is in the process of establishing a Curriculum Standards Review Board. The board will be comprised of subject matter experts from the emergency response, public health, and medical communities as well as representatives from organizations involved in the development of existing standards, for example, the National Fire Protection Association. This board will review existing and new courses to ensure they adhere to nationally accepted standards. Courses meeting the standards will be included in updated versions of the compendium.

The Department of Defense is committed to work with our interagency partners to establish effective national domestic programs and policies that will enhance the preparedness at all levels of government to respond to the awful consequences of a WMD attack. As has been discussed, DoD has a well-established pattern of federal assistance to state and local authorities in times of disaster and has leveraged that capability in developing the WMD DPP program. Although we can never be fully prepared to respond to all types of events in all locations, we have begun to lay the foundation for an integrated, across-the-board response—one that makes sense and one that is truly responsive to the needs of first responders. The continued partnership for WMD preparation among local, state and federal authorities will be essential to our success.

Thank you for the opportunity to participate in this important hearing.

ASSISTANT SECRETARY OF DEFENSE
WASHINGTON, DC 20301-1500

RESERVE AFFAIRS

16 JUL 1999

Honorable Tillie K. Fowler
Chairman
Subcommittee on Oversight,
Investigations and Emergency Management
Washington, DC 20515-6143

Dear Chairman Fowler:

Thank you for your recent letter with questions related to the issues discussed during the June 9, 1999, Subcommittee hearing. Enclosed please find the Department's response to your five follow-on questions.

If you have any additional questions, please contact my special assistant for military assistance to civil authorities, Ms. Ellen Embrey, at (703) 695-7305.

Sincerely,

Jim Schmitt for

Charles L. Cragin
Acting

Enclosure
As stated

Question 1. What is the schedule for the 10 RAID teams becoming operational?

- __Answer__: The teams are scheduled to be operational in January 2000.

Question 2. What is the schedule for the 44 "Lite" teams becoming operational?

- __Answer__: The 44 RAID Light teams will consist of traditional drilling National Guard soldiers and airmen. There is no specific operational timeline established for RAID Light teams. They are currently being organized and will be trained and equipped over the next several years as funding permits. Each state has been given the flexibility to customize its RAID Light training and equipment program using the RAID model as the baseline.

Question 3. What is the method and schedule for the Department of Defense (DoD) to evaluate the effectiveness of RAID teams?

- __Answer__: US Army Forces Command (FORSCOM) has been directed to provide training and readiness oversight for the RAID teams. Team members have completed almost all of their individual training and will be conducting new equipment training and an initial collective training exercise in July and August 1999. Following this training, teams will return to their home stations and continue to train as a unit. Each team will undergo their first no-notice validation exercise prior to the end of 1999, and annually thereafter.

Question 4. What is the status and schedule of any of the following activities that the Department of Defense may be performing:

- **an inventory and assessment of existing response capabilities and assets at the federal level;**

    - __Answer__: The DoD does not maintain an inventory or assessment of response capabilities of non-DoD agencies. The Department does maintain a listing of important DoD general assets in DoD Directive 5160.54, *Critical Asset Assurance Program*, and possesses various units capable of providing support to civil authorities responding to terrorists attacks that involve chemical, biological, radiological, nuclear, or high-explosive conventional weapons. (These weapons are commonly referred to as weapons of mass destruction (WMD).) The Department of Energy, the Environmental Protection Agency, and Department of Transportation are just a few of the other federal agencies possessing units capable of responding to WMD incidents. Collective information on those and other federal or state assets may be available through the Federal Emergency Management Agency (FEMA).

      Prior to the Nunn-Lugar-Domenici Act (NLD), the Army's Technical Escort Unit (TEU) and Marine Corps' Chemical-Biological Incident Response Force (CBIRF) existed and were capable of providing federal support for domestic WMD incidents. Additionally, all of the services possess explosive ordinance disposal (EOD) units at most major DoD installations.

      As a result of NLD, DoD specifically developed the Chemical-Biological Rapid Response Team (CB-RRT) as a federal asset to support domestic requirements for consequence management stemming from WMD incidents. Additionally, the National Guard was tasked to develop ten

Rapid Assessment and Initial Detection (RAID) teams to provide support to state authorities for WMD incidents. This was important since RAID teams were the first DoD-related units primarily designed as a state resource to employ in an event involving WMD.

Although the TEU, CBIRF, and most EOD units are capable of providing federal support to civil authorities for domestic or foreign WMD situations, their primary purpose is to respond to DoD wartime requirements or to support unique ordnance handling for service-specific mission areas. This capability includes handling nuclear, chemical or biological waste, weapons and materials. Conversely, the CB-RRT and RAID teams were specifically designed for domestic support and not for DoD wartime operations. To clarify the structure, capability, and purpose of those teams or units, the following brief synopsis is provided.

TEU: In 1943 the Army established the TEU with a mission to conduct field sampling, identification, verification, monitoring, recovery, decontamination, transportation, exploitation, escort, and mitigation of hazards associated with weaponized and non-weaponized chemical and biological materials. TEU also provides chemical/biological response teams (CBRT) to government agencies in support of national and international counterproliferation programs. The TEU includes a total of about 85 military and 67 civilian personnel. By design, these personnel can be formed into various sized response elements, but are generally considered as four teams.

CBIRF: The Marine Corps' CBIRF is a self-contained 350 person unit located in North Carolina and is deployable on a world-wide basis. It is trained as a rapid response force capable of threat identification, casualty extraction, personnel decontamination and medical triage, treatment, and stabilization for terrorist-initiated chemical and biological attacks. It can support wartime requirements or could support US or foreign domestic events involving WMD.

EOD Units: The Army's 52nd Ordnance Group is one of the more well-known EOD groups and has units located throughout the country capable of EOD support. EOD units attached to the 52nd Ordnance Group can be deployed to world-wide locations to defeat or mitigate hazards from conventional, nuclear, or chemical devices. Other service-specific EOD units handle materials or explosives commonly processed within certain mission areas. Major DoD installations usually have EOD units assigned and trained to handle some types of weaponized chemical, biological or explosive munitions. However, their primary mission is to support warfighting operations or munitions handling.

CB-RRT: The CB-RRT, which resulted from the NLD legislation, is unique since it is a skeletal organization designed specifically to support the lead federal agency (LFA) for a civilian crisis resulting from a WMD event. It draws DoD assets to assist in the detection, neutralization, containment, dismantlement and disposal of WMD articles or related hazardous materials, and it assists first responders in dealing with potential WMD consequences.

Unlike the CBIRF, the CB-RRT must task and operationally control, or establish liaison with, a variety of units from within the DoD (i.e., TEU, RAIDS, CBIRF, the 52d Ordnance Group, logistics units, transportation units, military police, and military medical units) to accomplish the response mission. For practical purposes, the CB-RRT is the standing framework that combines existing DoD assets when needed for a domestic response force.

The Army's Soldier and Biological Chemical Command (SBCCOM) at Fort Meade provides the core CB-RRT architecture. SBCCOM also makes available assets from the Edgewood Chemical and Biological Center (ECBC), the US Army Forensic Analytical Center (USAFAC), and the Chemical Activity Commands (CACs) as part of the CB-RRT response capabilities. These capabilities represent specific units assigned to the CB-RRT by the Director of Military Support (DOMS) for civil support requirements.

Other resources available to CB-RRT include the Army's Medical Research Institute of Infectious Disease (USAMRIID), Medical Research Institute of Chemical Defense (USAMRICD), and the Center for Health Promotion and Preventive Medicine (USACHPPM). Available also are the Service's special medical response teams from the six regional medical commands (RMC) and resources from the Naval Research Laboratory (NRL), Naval Medical Research Institute (NMRI), Naval Environmental Health Center (NEHC), and the NEHC's Naval Environmental and Preventive Medicine Units.

RAID: RAID teams are part of the National Guard and are intended to be primarily a state asset readily available to a governor to assist state and local civil authorities. Because of the unique nature of the National Guard as a state and federal force, RAID teams may also be employed as a federal asset. However, their specific mission is to assess WMD situations, identify WMD agents, recommend appropriate courses of action to civil authorities to mitigate damage from the effects of the WMD, and to assist civil and military authorities in facilitating the employment of federal military support. RAID teams will be dual-hatted as both a state and federal military resource. Again, they are primarily intended to be employed at the direction of a governor, but may be federalized to support DoD requirements since all National Guard units are inherently part of the DoD. RAID teams are scheduled to be operationally capable in January 2000.

In summary, the DoD can utilize a variety of technical organizations or medical units to identify and respond to nuclear, chemical and biological WMD events or to mitigate the potential health consequences resulting from the use of a WMD. However, the primary purpose of most of these units is to enhance DoD wartime operations or to support their respective service in handling chemical, biological, nuclear or conventional munitions, materials, waste or devices. Only the DoD's CB-RRT and RAID teams have been specially designated by law to support civil authorities in responding to WMD events. All other DoD units are designed to support wartime missions but could be employed, if required, to support civil authorities.

- **an inventory and assessment of existing response capabilities and assets at the state and local level;**

  - **Answer**: The DoD provides support to state and local officials when a governor formally requests assistance from the President, but it does not maintain an inventory of existing response capabilities and assets at the state or local level. State emergency management agencies usually maintain listings of state and local assets within their region. FEMA might also maintain similar data, but the DoD does not maintain this level of detail on capabilities and assets at the state and local level. The DoD's Director of Military Support (DOMS) acts as the liaison to interact with

FEMA and other governmental agencies whenever the DoD is requested to support civilian catastrophes.

In spite of the fact that many areas of the country have Hazardous Materials (response) Teams (HAZMATS), RAID teams are being trained to recognize weaponized chemical, biological, and nuclear materials used by terrorists. They are also being trained to provide civilian authorities with viable recommendations on mitigating the consequences resulting from the exposure of WMD materials on the civilian population. Consequently, RAID teams are the only state asset that the DoD is monitoring as a state or local asset designated for WMD consequence management.

- **an identification of response needs at the federal level;**

  - **Answer**: DoD is not specifically responsible to identify response needs at the federal level; however, through the Office of the Assistant Secretary of Defense for Special Operations and Low Intensity Conflict (OASD (SO/LIC)), DoD is lending support to the FBI and FEMA in pursuing the development of a capacity and sustainment study designed to examine the potential requirements for national teams and resources for WMD consequence management for domestic crises. Some of the issues they will be assessing include ascertaining our policy for and funding of national teams and assets for WMD response, and determining the number and capability of such teams.

    Specifically, OASD(SO/LIC) is examining the utility of a computer modeling system designed to evaluate a spectrum of terrorist incidents involving six types of WMD: nuclear, radiological, chemical (both persistent and non-persistent agents) and biological (contagious and non-contagious agents). Computer models would be generated to evaluate the scope of personnel and collateral damage incurred in an urban environment by the use of these weapons. A generic urban environment will be used initially to avoid singling out any particular city in the U.S. An initial operative assumption for the study is that this city would have benefited from participation in the NLD Domestic Preparedness Program for 120 of America's larger metropolitan areas.

    The six types of WMD will be examined individually in scenarios designed to illustrate their unique characteristics including: resultant human casualties; damage to urban infrastructure; degree and persistence of contamination including compromises to water, food supplies, agricultural assets, health care facilities, and communication and transportation systems. The analysis will also consider diffusion patterns indicating the pervasiveness of the WMD attack; effects of meteorological conditions on the attack; impact on population mobility; and in the case of biological weapons use, timelines for the onset of certain symptoms based on transmissibility and incubation rate in humans.

    Throughout the course of this study, these weapons characteristics will be amplified to examine worst case scenarios and the corresponding national effort to mitigate single incidents as well as simultaneous terrorist incidents. By varying the parameters used to construct the model, mathematical data will be generated which will serve to define the requirements for response resources for WMD consequence management.

OASD(SO/LIC) is also working to ascertain the extent of modeling conducted to date for consequence management (CM) response. The Defense Threat Reduction Agency, the Central Intelligence Agency (CIA), USAMRIID, SBCCOM, the National Ground Intelligence Center and the Department of Health and Human Services (DHHS) are also being queried for details on previous computer modeling efforts in CM response. Extensive modeling work has already been undertaken by the National Ground Intelligence Center which has developed a Joint Assessment of Catastrophic Events (JACE) program for DOMS. The JACE effort is based on leveraging a set of targeting tools and modeling software in a web-based collaborative environment for use by military and civilian responders in the event of a WMD attack. One desired outcome would be that a virtual city construct could be established for the generation of consequence management requirements for the six aforementioned WMD scenarios.

OASD(SOLIC) will be working with the CIA and Defense Intelligence Agency (DIA) to prioritize the WMD scenarios. The results of this study will be provided to FEMA in order to determine and coordinate eventual Federal Response Plan (FRP) support. FEMA will coordinate with DHHS, the FBI and other FRP contributors. Following completion of the initial modeling scenarios, additional scenarios will be examined to broaden the database. This study is an ongoing process designed to minimize the element of surprise for national policy makers and bolster the national response capability.

- **an identification of response needs at the state and local level; and**

  - __Answer__: As previously stated, the DoD is supporting the FBI and FEMA in pursuing the development of a generic urban environment computer model study designed to examine the potential requirements for national teams for hypothetical urban WMD crises. However, the DoD is not specifically focusing on the needs of a particular state, city or local level of government. Instead, the DoD is supporting the joint FEMA-FBI capacity study to help establish rough estimates for state, county and city managers so that the information could be employed as a means to establish a baseline of requirements for potential CM catastrophes.

- **a written threat and risk assessment of our cities or regions of the United States regarding a terrorist attack.**

  - __Answer__: The FBI is the federal agency responsible for threat and risk assessment of our cities and regions of the United States regarding a terrorist attack. Additionally, you may be able to gain this information from some state, county, or city emergency management offices or law enforcement agencies. The DoD does not normally conduct standing or recurring assessments of this nature.

Question 5. What annual goals and measures developed by Department of Defense pursuant to the Government Performance and Results Act (GPRA) relate to preparedness against terrorist attack?

- __Answer:__ The goals presented in the Department's fiscal year 2000 GPRA Performance Plan support the National Security Strategy and the Department's mission and vision statements. They were selected because they demonstrate how senior decisionmakers judge the Department's progress toward meeting the long-term strategic goals of the Quadrennial Defense Review (QDR). As such, the measures and indicators that define these goals evaluate performance at an executive level. They focus on overarching measures and indicators that demonstrate the ability of the Department to carry out a range of missions, from counterterrorism to fighting a major theater war.

For example, the QDR commits the Department to "*shape* the international environment and *respond* to the full spectrum of crises by providing appropriately sized, positioned, and mobile forces." Because the QDR serves as the Department's GPRA Strategic Plan, this commitment is expressed as the GPRA corporate goal: "Shape and Respond." Further, the QDR suggests the level of performance that best satisfies this corporate goal is a military force capable of multiple missions – a force that is proficient in core warfighting competencies as well as able to meet asymmetric challenges, including terrorism. Therefore, the Department's GPRA plan selected annual performance goals that protect the structure and readiness for such a multi-mission force.

Following this logic, the annual performance goal most closely associated with responding to asymmetric threats such as terrorism is Goal 1.2: "Maintain ready forces and ensure they have the training necessary to provide the United States with the ability to shape the international environment and respond to the full range of crises." In addition, the plan's force level metrics (1.2.1, 1.2.3, 1.2.5, and 1.2.7) capture the resources required to maintain military forces ready to respond to the full range of crises, including asymmetric attack. The plan's training metrics (1.2.17, 1.2.18, and 1.2.19) set goals for all training by type of service and type of combat system (aircraft, combat and support vehicles, and ships), including counterterrorism and counter-WMD training for appropriate units. In addition, the plan outlines cross-cutting efforts to defend against terrorism (including potential terrorist use of weapons of mass destruction), and describes how the Department provides technical support to the FBI and FEMA to guard against physical or cyber attack.

Preparedness for Terrorism Response

Statement by Chief John M. Eversole

presented to

Subcommittee on Oversight, Investigations, and Emergency Management

of the

Committee on Transportation and Infrastructure

U.S. House of Representatives

June 9, 1999

International Association of Fire Chiefs
4025 Fair Ridge Drive · Fairfax, VA 22033-2868
(703) 273-0911 · FAX (703) 273-9363

I am Chief John M. Eversole of the Chicago Fire Department. I also appear before you as Chair of the Hazardous Materials Committee of the International Association of Fire Chiefs (IAFC).

The issue of domestic terrorism is one in which America's fire departments have a vital interest. Violence perpetrated against our citizens for political purposes, national, international or otherwise, will be suffered locally. As the primary provider of emergency life safety services, fire fighters will be first on the scene of any act of terrorism, saving lives and mitigating damage. This was true in the minutes following the 1993 World Trade Center bombing and the 1995 bombing of the Alfred P. Murrah federal building in Oklahoma City. It has been so at countless incidents less notorious. So it will be in the future.

There are two distinct areas of federal counterterrorism efforts that should be addressed. First, programs designed to support, prior to an incident, local emergency services personnel who will be first on the scene and second, the operational role of federal agencies in the wake of an attack. I will address the pre-incident support role first.

The Nunn/Lugar/Domenici amendment to the 1997 Defense Authorization and the Antiterrorism and Effective Death Penalty Act of 1996 began federal efforts to help better prepare local fire, police and emergency services agencies for the possibility of terrorism involving chemical, biological, radiological and conventional weapons. The IAFC was involved in the development of both these laws and continues to work with the Departments of Defense and Justice in their administration.

The Antiterrorism Act authorized a $5 million appropriation to train metropolitan fire fighters in terrorism response. Designated by the Attorney General to administer this law, the Office of Justice Programs (OJP) provided four jurisdictions with demonstration grants and, importantly, worked with the National Fire Academy in the development of awareness-level training curriculum that has been available nationwide for two years. A train-the-trainer approach was used for both cost savings and an efficient way to reach as many fire fighters as possible. Tens of thousands have received training based on these materials. Successfully mitigating a terrorist incident is incumbent upon early identification. Awareness-level training is vital and should continue to be provided.

OJP also created, at the direction of Congress, a National Domestic Preparedness Consortium comprised of Louisiana State University, the New Mexico Institute for Mining and Manufacturing, Texas A&M University and the Nevada Test Site's explosive ordinance facilities. The Justice Department also took control of the U.S. Army's chemical weapons training facilities at Ft. McClellan, Alabama, and designated this facility as the National Domestic Preparedness Center. Managers at these facilities have actively sought out expertise from the fire service and have demonstrated a willingness and ability to respond to constructive criticism of their programs. We support expedited access to the Consortium's facilities for as many local emergency services personnel as possible. It is essential that local emergency response agencies have significant input and oversight of program development. It is important that what is being taught is not duplicative and that training meets our needs.

Both the Defense and Justice Department programs have been criticized for focusing on only 120 of our largest cities. The 120 city approach, though arbitrary, allowed both agencies to focus on jurisdictions that collectively contain a majority of our population and provided a goal that was attainable with limited resources. We have stated repeatedly that this goal needs to be broadened. We believe that fire fighters in all communities need some level of basic training. The awareness level-training developed by OJP in concert with the National Fire Academy is a good starting point. This training needs to be made available nationwide. Sufficient funding should be provided by Congress to ensure its availability and delivery everywhere.

The IAFC believes that the enhancement of existing local capabilities is the wisest, most cost-effective course to follow in preparing for "weapons of mass destruction" terrorism. It is our experience that not only will we be the first responders on the scene, but we will be the largest supplier of personnel and equipment throughout the incident. Fire department hazardous materials response teams deal with spills and accidental releases of highly toxic chemicals on a regular basis. This is the case across the country. Additional training in safely containing chemical, biological or radiological agents is a high priority for the fire service.

The emphasis to-date has been on "WMD" terrorism. The importance of preparing for this risk cannot be understated. However, we should not forget that bombing attacks remain the most common tool of terrorists both domestically and abroad. We should not overlook this fact in our preparedness plans. And we do realize that the National Domestic Preparedness Consortium's New Mexico program addresses this problem.

Assistance in training and equipping the fire service to respond effectively to terrorism is what is most needed from the federal government. The Nunn/Lugar and Antiterrorism Act programs are both beneficial. Both can be improved. Current plans to transfer the Nunn/Lugar program to the Justice Department make sense. Our experience with the Department of Justice has been generally positive. The Department of Justice has a demonstrated ability to deliver support programs to local agencies.

The second role of the federal government is that of operational response. There are a host of operational response assets at the federal level. Some of these are: the U.S. Marine Corps' CBIRF, the National Guard's proposed RAID teams, the FBI's HMRU, the U.S. Army's Technical Escort Unit, the EPA's Emergency Response Teams and the Department of Energy's Radiological Emergency Response teams. The list goes on. Most, if not all, of these entities have impressive capabilities and are staffed and led by intelligent, competent professionals. We respect their abilities. But we cannot afford duplicative teams. We must have one good federal operational unit with the right capabilities.

One final point regarding federal response teams. America's fire service has 350-plus years of experience in mitigating disasters and emergencies of all kinds, including acts of terrorism. We have learned that federal assets are only as good as their proximity to the scene of an incident.

I turn now to the coordination of these support programs and to the operational coordination of federal agencies in response to an actual incident. The Attorney General has created a National Domestic Preparedness Office (NDPO) within the Federal Bureau of Investigation to serve as a contact point for local agencies and to facilitate coordination of planning, training and operational programs. The IAFC has worked with the FBI for several years and applauds this effort.

As I have said, numerous federal agencies have response capabilities though it is currently unclear how these units will work together. We feel that the proposed State and Local Advisory Group of the NDPO could be critical in order to address the need for a well-coordinated response to any incident of terrorism. This group could also provide the necessary input to deliver appropriate support and training so needed by local emergency response agencies. Although we support, in concept, an NDPO, we reserve judgement until such time as the NDPO has proven itself.

Thank you for inviting me to testify today. I am happy to answer any questions you may have.

Statement of
Mark Gebicke
Director, National Security and Preparedness Issues
General Accounting Office

Before the
Subcommittee on Oversight & Investigations
House Transportation and Infrastructure Committee
U.S. House of Representatives

Hearing on Preparedness Against Terrorist Attacks

June 9, 1999

Madam Chairman and Members of the Subcommittee:

I am pleased to be here to discuss our prior work and observations on federal efforts to combat terrorism, especially those to prepare for and respond to terrorist attacks involving chemical, biological, radiological, or nuclear (CBRN) weapons or devices. As you know, the President's fiscal year 2000 budget requested about $10 billion to combat terrorism. According to the Office of Management and Budget (OMB), about $1.4 billion of that amount was for dealing with "weapons of mass destruction." Over the past 3 years we have evaluated and reported on a number of issues concerning federal programs and activities to combat terrorism. A list of related GAO reports and testimonies is attached to this statement.

My testimony will focus on three issues. First, I will briefly describe the foreign- and domestic-origin terrorism threats, as we understand them from intelligence analyses, and discuss some issues surrounding the emerging threat of CBRN terrorism. Second, I will provide our observations on the growth in federal programs to provide training and equipment to local "first responders"—police, fire, and emergency medical services—and the expansion of federal response elements and teams to deal with a possible CBRN terrorist attack. Finally, I will discuss some steps the executive branch has taken to better manage federal efforts to combat terrorism and some opportunities we see for additional focus and direction.

SUMMARY

U.S. intelligence agencies continuously assess both the foreign and domestic terrorist threat to the United States and note that conventional explosives and firearms continue to be the weapons of choice for terrorists. Terrorists are less likely to use chemical and biological weapons than conventional explosives, although the possibility that they may use chemical and biological materials may increase over the next decade, according to intelligence agencies. Agency officials have noted that terrorist use of nuclear weapons is the least likely scenario, although the consequences could be disastrous. Although the intelligence agencies agree on these matters, we have observed many conflicting statements and views in public documents and testimony about the CBRN terrorism threat. In addition, there is an apparent disconnect between the intelligence agencies' judgments and the focus of certain programs.

Since 1996, the number of federal programs and initiatives to combat terrorism have grown significantly. According to the Office of Management and Budget, funding has also increased from about $6.5 billion in fiscal year 1998 to about $10 billion requested for fiscal year 2000. At the same time that the federal government has created several potentially overlapping programs to train and equip local first responders to prepare for possible CBRN terrorist attacks, federal agencies have also expanded the number of federal response teams, capabilities, and assets.

The executive branch has taken some important steps toward improving the way it manages and coordinates the growing, complex array of agencies, offices, programs, activities, and capabilities. For example, OMB has issued two governmentwide reports—one in 1998 and one in 1999—on funding levels and programs to combat terrorism. In addition, in December 1998, the Attorney General issued a classified 5-year interagency plan on counterterrorism and technology. The Attorney General is also establishing a National Domestic Preparedness Office at the Federal Bureau of Investigation (FBI) to try to reduce state and local confusion over the many federal training and equipment programs to help them prepare for terrorist incidents involving CBRN weapons. While these are important positive steps, we see opportunities to improve the focus and direction of federal programs and activities to combat terrorism. For example, a governmentwide strategy that includes a defined end-state and priorities is needed, along with soundly established program requirements based on assessments of the threat and risk of terrorist attack. In addition, a comprehensive inventory of existing federal, state, and local capabilities that could be leveraged or built upon is warranted before adding or expanding federal response assets. Without these fundamental program elements, there can be little or no assurance that the nation is focusing its investments in the right programs and in the right amounts and that programs are efficiently and effectively designed and implemented.

BACKGROUND

Under Presidential Decision Directive 39 (June 1995) federal efforts to combat terrorism are organized along a lead agency concept. The Department of Justice, through the FBI, is the lead federal agency for crisis management of domestic terrorist incidents and for pursuing, arresting, and prosecuting the terrorists. For managing the consequences of domestic terrorist incidents, state and local authorities are primarily responsible. If federal assistance is requested, the Federal Emergency Management Agency (FEMA) is the lead federal agency for consequence management. FEMA coordinates this federal support through the Federal Response Plan, which outlines the roles, responsibilities, and emergency support functions of various federal agencies for consequence management. The National Coordinator for Security, Infrastructure Protection, and Counterterrorism at the National Security Council is charged with coordinating the broad variety of relevant policies and programs including such areas as counterterrorism, preparedness, and consequence management for CBRN terrorist incidents.

THE FOREIGN- AND DOMESTIC-ORIGIN TERRORISM THREAT IN THE UNITED STATES

Terrorist bombings of the World Trade Center in New York City in 1993 and the federal building in Oklahoma City in 1995 have elevated concerns about terrorism in the United States. Previously, the focus of U.S. policy and legislation had been on international terrorism abroad and airline hijacking. Intelligence agencies continuously assess the foreign and domestic terrorist threats to the United States. The U.S. foreign intelligence community, which includes the Central Intelligence Agency and others, monitors the foreign-origin terrorist threat to the United States. In addition, the FBI gathers intelligence and assesses the threat posed by domestic sources of terrorism.

What is important about these assessments is the very critical distinction between what is conceivable or possible and what is likely in terms of the threat of terrorist attack. While concerns about terrorist use of CBRN weapons were heightened by an apocalyptic sect's use of a nerve agent in the Tokyo subway in

1995, terrorists are still reportedly more likely to use conventional weapons. According to the U.S. intelligence community, conventional explosives and firearms continue to be the weapons of choice for terrorists, at least partly because chemical and biological agents are more difficult to weaponize and the results are unpredictable.

On average, from 1992 through 1998, there were fewer than four terrorist incidents in the United States each year, according to FBI statistics. Figure 1 provides FBI data on the number of terrorist incidents in the United States during the 1992-98 period, none of which were CBRN attacks.

Figure 1: Terrorist Incidents in the United States, 1992-98

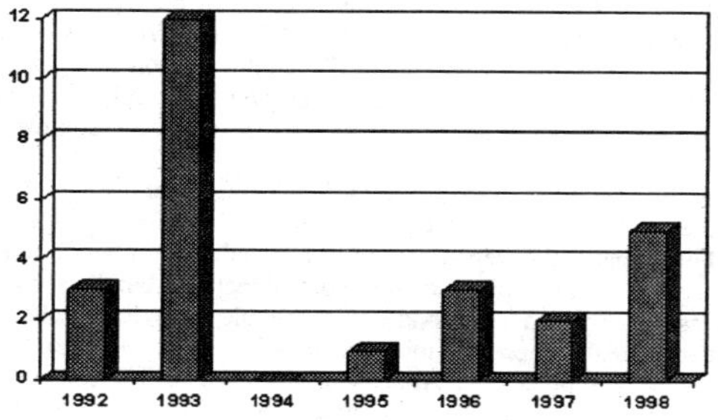

Source: FBI

The intelligence community reports that some foreign-origin groups and individuals of concern are showing an increasing interest in using chemical and biological materials. The FBI also reports an increasing number of domestic cases involving U.S. persons attempting or threatening to use such materials. Agency officials have noted that, although the consequences could be disastrous, the terrorist use of nuclear weapons is the least likely scenario.

Issues Surrounding the Emerging

CBRN Terrorism Threat

Statements made in testimony before the Congress and in the press by various officials on the issue of making and delivering a terrorist chemical or biological weapon sometimes contrast sharply. On the one hand, some statements suggest that developing a chemical or biological weapon can be relatively easy. For example, in 1996, the Central Intelligence Agency Director testified that chemical and biological weapons can be produced with relative ease in simple laboratories, and in 1997, the Central Intelligence Agency Director said that "delivery and dispersal techniques also are effective and relatively easy to develop."

Similarly, an article by former senior intelligence and defense officials noted that chemical and biological agents can be produced by graduate students or laboratory technicians and that general recipes are readily available on the internet.

On the other hand, some statements suggest that there are considerable difficulties associated with successfully developing and delivering a chemical or biological weapon. For example, the former Deputy Commander of the Army's Medical Research and Materiel Command testified in 1998 that "an effective, mass-casualty producing attack on our citizens would require either a fairly large, very technically competent, well-funded terrorist program or state sponsorship." More recently, in March 1999, the Special Assistant to the Director of Central Intelligence for Nonproliferation testified that "the preparation and effective use of biological weapons by both potentially hostile states and by non-state actors, including terrorists, is harder than some popular literature seems to suggest."

We are reviewing the scientific and practical feasibility of the terrorist chemical and biological threat for the Senate Committee on Veterans Affairs; the Ranking Member of the House Armed Services Committee; and the House Government Reform Committee's Subcommittee on National Security, Veterans Affairs, and International Relations. Specifically, we are examining the ease or difficulty for a non-state actor to successfully obtain chemical and biological agents, process the materials, and make and deliver chemical and biological weapons that can cause mass casualties. We plan to issue our report later this summer.

We have also observed a disconnect between intelligence agencies' judgments about the more likely terrorist threats—particularly the chemical and biological terrorist threat--and certain domestic preparedness program initiatives. For example, the Department of Health and Human Services' (HHS) fiscal year 1999 budget amendment proposal for its bioterrorism initiative included building--for the first time--a civilian stockpile of antidotes and vaccines to respond to a large-scale biological or chemical attack and expanding the National Institutes of Health's research into related vaccines and therapies. Specifically, the Omnibus Consolidated and Emergency Supplemental Appropriations Act (P.L. 105-277) included $51 million for the Centers of Disease Control and Prevention to begin developing a pharmaceutical and vaccine stockpile for civilian populations.

HHS' legislatively required operating plan discusses several chemical and biological agents selected for its stockpiling initiatives. These agents were selected because of their ability to affect large numbers of people (create mass casualties) and tax the medical system. We observed that several of the items in HHS' plan did not match individual intelligence agencies' judgments, as explained to us, on the more likely chemical or biological agents a terrorist group or individual might use. HHS had not documented its decision making process for selecting the specific vaccines, antidotes, and other medicines cited in its plan. Thus, it was unclear to us whether and to what extent intelligence agencies' official, written threat analyses were used in the process to develop the list of chemical and biological terrorist threat agents against which the nation should stockpile. Further, we have not seen any evidence that HHS' process incorporated the many disciplines of knowledge and expertise or divergent thinking that is warranted to establish sound requirements to prepare for such a threat and focus on appropriate medical preparedness countermeasures.

GROWTH IN FEDERAL FUNDING,

PROGRAMS, AND INITIATIVES

Federal funding of efforts to combat terrorism has increased rapidly. According to the Office of Management and Budget, funding to combat terrorism has increased from about $6.5 billion in fiscal year 1998 to about $10 billion requested for fiscal year 2000. Overall, the number of agencies, offices, and initiatives to combat terrorism has also grown substantially. Specifically, since 1996, we have observed growth in federal funding and programs to provide training and equipment to local first responders and, concurrently, growth and potential overlap in federal response elements and teams to deal with a possible CBRN terrorist attack. The federal response elements and assets have been established to support state or local incident commanders to manage the consequences of a possible CBRN terrorist attack.

Proliferation of Federal Programs to

Train and Equip First Responders

We have observed a proliferation of programs and initiatives across several agencies to provide training and/or equipment to local first responders for dealing with the consequences of a CBRN terrorist attack. On the surface, it appears to us that there is potential for duplication and overlap among these programs. The fiscal year 2000 budget request proposed $611 million for training, equipping, and exercising cities' first responders in preparation for a potential terrorist attack and for strengthening public health infrastructure. Table 1 summarizes some aspects of selected federal training and/or equipment programs available to state and local agencies to build or enhance their CBRN response capabilities.

Table 1: Selected Federal CBRN Consequence Management Training and/or Equipment Programs

Agency	What program provides	Target audience
Department of Defense (DOD)	Training: CBRN response with focus on chemical, biological, and nuclear. Equipment: Provides each city up to $300,000 in equipment on 5-year loan.	Police, fire, hazardous materials technicians, and medical and emergency management responders in the 120 most populous cities.
Department of Justice	Training: Explosives, incendiary, chemical, and biological (not radiological or nuclear) response. Equipment: Provides equipment grants.	Police, fire, hazardous materials, and medical and emergency management responders in the 120 largest urban jurisdictions.
FEMA	Training: Emergency management and hazardous materials response, including those related to terrorist incidents.	Fire, medical, hazardous materials technicians, and other emergency responders.
Department of Energy	Training: Nuclear and radiological response in emergencies.	Responders in communities close to nuclear facilities.
Environmental Protection Agency (EPA)	Training: Chemical, biological, and radiological hazardous materials response, with	Federal, state, and local hazardous materials technicians.

	new focus on terrorist "weapons of mass destruction" incidents.	
HHS	Equipment: Contract grants include funds for equipment and items for medical response to CBRN incident.	Emergency medical responders in 27 cities that also participate in DOD's Domestic Preparedness Program.
Department of Veterans Affairs (VA)	Training: CBRN incident, with focus on medical response. Training to be provided under contract with HHS.	1,100 nonfederal National Disaster Medical System hospital staffs.

Source: GAO

Further information on these federal programs and activities is at appendix I.

Some local officials we spoke with during our examination of DOD's Domestic Preparedness Program viewed the growing number of CBRN consequence management training programs as an indication of a fragmented and possibly wasteful federal approach toward combating terrorism. Similarly, multiple equipment programs were causing frustration and confusion at the local level and were resulting in further complaints that the federal government is unfocused and has no coordinated plan or defined end-state for domestic preparedness. For example, in the Domestic Preparedness Program, the separation of the DOD and HHS equipment packages required local officials to deal with two federal agencies' differing requirements and procedures. Since the HHS equipment program is offered through a contract with unmatched federal funds, the cities had to meet certain requirements, including developing a concept of operations plan for Metropolitan Medical Response Systems that fits into a local area's overall medical response system. The DOD equipment loan program required a different process. Other equipment initiatives, such as the Department of Justice equipment grant program, could add to the local government officials' perception of an unfocused federal strategy.

Growth in Federal CBRN

Response Elements

At the same time federal training and equipment programs for first responders has grown, the number of federal response elements that can deal with various aspects of managing the consequences of a CBRN terrorist attack has also expanded and increased. Individual agencies' initiatives include adding teams or capabilities that can identify and analyze various chemical and biological materials or agents; contain or handle the weapon, device, or area of an incident; and provide medical support or response for dealing with potential casualties of an incident. We have pointed out that the growth in these capabilities and assets has not been based on soundly established requirements or a comprehensive inventory of existing federal, state, and local assets that could be leveraged. State and local officials have raised concerns about the increasing number of federal response elements being formed. In our view, the emergence of more federal response elements and capabilities will increase the challenge for the federal government to provide a well-coordinated response in support of a state or local incident commander.

DOD has established several new response elements in addition to those that have been or would have been called upon in the past to respond to potentially dangerous chemical or biological threats or incidents. Among the pre-existing response assets are the Army's Technical Escort Unit, which has four teams in two U.S. locations and the Army's 52nd Explosives Ordnance Disposal Group, which includes many units located throughout the country and has personnel specially trained to respond to CBRN incidents. In 1996, the Marine Corps created the Chemical Biological Incident Response Force located at Camp LeJeune, N.C., to provide a medical and decontamination response to CBRN incidents. In addition, the Nunn-Lugar-Domenici Act required DOD to establish a Chemical-Biological Rapid Response Team for domestic incidents.

More recently, DOD has created National Guard Rapid Assessment and Initial Detection (RAID) teams in 10 states to respond to CBRN incidents. Potentially, up to 54 RAID teams are planned. The RAID teams' mission is to provide assistance to local and state authorities in the event of an incident involving chemical, biological, nuclear, or radiological weapons. They are to (1) help assess the situation, (2) advise civilian responders as to appropriate actions, and (3) facilitate the identification and movement of federal military assets to the incident scene. We reviewed the roles and missions of the RAID teams and expect to release a report this month.

As mentioned earlier, the Department of Health and Human Services (HHS) has established Metropolitan Medical Response Systems with trained and equipped local emergency teams in 27 cities that also participate in the DOD-led Nunn-Lugar-Domenici Domestic Preparedness Program. HHS requested fiscal year 2000 funding to include 25 more cities in its program. In addition to the 27 locally-based medical response teams (with more to be established), HHS has established four specialized National Medical Response Teams, three of which are deployable in the event of a terrorist attack involving a chemical or biological weapon. These 27 Metropolitan Medical Response Systems and 4 National Medical Response Teams are in addition to HHS' 24 Disaster Medical Assistance Teams that deploy to provide medical support for any type of disaster, including terrorism. HHS is further expanding its response capabilities by creating a national stockpile of millions of doses of vaccines, antidotes for chemical agents, antibiotics for other diseases, and respirators.

Another federal response element that appears to be growing is federal laboratories with capability to analyze chemical and biological agents. The Army, the Navy, and the Centers for Disease Control and Prevention have had laboratory capabilities to analyze chemical and biological agents. In addition, HHS has plans to establish regional laboratories, and the FBI is establishing a mobile laboratory capability. Both the FBI and EPA have forensic laboratories, although there are some differences in capabilities, and the FBI is looking into using existing facilities rather than creating a specialized laboratory for CBRN cases.

SOME STEPS TAKEN, BUT OPPORTUNITIES REMAIN TO IMPROVE MANAGEMENT OF CROSSCUTTING PROGRAMS

The executive branch has taken a number of important steps to improve management and coordination of programs to combat terrorism. Nevertheless, we have pointed out several areas in which fundamental program elements are missing while program growth continues.

Steps Taken Toward

Improved Management

And Coordination

I will highlight four executive branch efforts that represent important steps toward improved management and coordination of the growing programs and activities to combat terrorism. First, OMB has started to track spending by federal agencies to combat terrorism. In December 1997, we reported that key federal agencies with responsibilities to combat terrorism spent about $6.7 billion in fiscal year 1997 for unclassified terrorism-related activities and programs and noted that precise funding information was unavailable for various reasons. That report led to legislation (National Defense Authorization Act for fiscal year 1998) requiring OMB to establish a system for collecting and reporting information on executive agencies' spending and budgets for combating terrorism. We believe that the OMB reports on governmentwide spending and budgeting to combat terrorism are a significant step toward improved management and coordination of the complex and rapidly growing programs and activities. For the first time, the executive branch and Congress have strategic oversight of the magnitude and direction of federal funding for this priority national security and law enforcement concern. The 1999 report provided additional analysis and more detailed information than the 1998 report on budgeting for programs to deal with CBRN weapons. For example, the 1999 OMB report identified the funding (budget authority) for the CBRN portion of combating terrorism to be about $1.23 billion in fiscal year 1999 and $1.39 billion in the fiscal year 2000 budget request.

Nevertheless, OMB officials told us, as we noted in our December 1997 report, that a critical piece of the budget and spending picture is missing--threat and risk assessments that would suggest priorities and appropriate countermeasures. These officials noted—and we agree--that risk assessment is key to (1) knowing whether enough or too much is being spent, (2) judging whether the right programs are being funded, and (3) determining whether apparent duplication is good or bad. We have not fully evaluated the processes or methodologies the executive branch agencies used to derive the information in the 1998 and 1999 OMB reports. As a result, we cannot comment on whether or to what extent the reports reflect the best possible estimate of costs associated with programs and activities to combat terrorism. The reports, however, do not clearly or explicitly describe any established priorities or duplication of efforts as called for in the legislation.

A second step toward improved interagency management and coordination was the Attorney General's December 1998, classified 5-year interagency plan on counterterrorism and technology crime. The Conference Committee Report accompanying the 1998 Appropriations Act for the Departments of Commerce, Justice, and State, the Judiciary, and Related Agencies required the Attorney General to develop the plan in coordination with several agencies. The plan includes goals, objectives, and performance indicators and recommends that specific actions be taken to resolve interagency problems and issues it identified and assigns relative priorities to the actions. The classified plan represents a substantial interagency effort and was developed and coordinated with 15 federal agencies with counterterrorism roles. The plan, however, generally does not link its recommended actions and priorities to budget resources, although it states that the agencies hope to improve the link between the plan and resources in subsequent updates. The plan also does not have a clearly defined end-state that would be useful to establish requirements and priorities.

A third step was the Attorney General's proposed establishment of a National Domestic Preparedness Office to coordinate the programs and other federal support for state and local governments. The purpose of the office is to coordinate Justice programs with those of other federal agencies to enable state and local first responders to establish and maintain a crisis and consequence management infrastructure capable of responding to a conventional and nonconventional terrorist attack. The office, under the leadership of the

Federal Bureau of Investigation, would address planning, training, equipment, exercises, research and development, intelligence and information sharing, and health and medical service needs at the federal, state, and local levels. The office has commissioned a local, state, and federal interagency board to establish, maintain, and update a standardized equipment list for use by the interagency community in preparing state and local jurisdictions to respond to a terrorist incident involving a weapon of mass destruction. The office is intended to reduce state and local confusion over the multitude of federal training and equipment programs and response capabilities by providing "one stop shopping" for state and local agencies. We understand that this office has not been formally approved.

Finally, in Presidential Decision Directive 62, issued in May 1998, the President designated a National Coordinator for Security, Infrastructure Protection, and Counterterrorism. While this coordinator is not to direct agencies' activities, he is responsible for integrating the government's policies and programs on unconventional threats to the homeland and Americans abroad, including terrorism. He is also to provide advice in the context of the annual process regarding the budgets for counterterrorism. We understand he has established a number of interagency working groups, but we have been unable to obtain any further information on these groups' responsibilities and accomplishments.

Opportunities to Enhance

Program Focus and Direction

Notwithstanding these important steps taken by the executive branch, we continue to see opportunities to better focus the nation's investments and efforts to combat terrorism. In November1998, we concluded that the many federal CBRN consequence management training, equipment, and response initiatives could benefit from a coordinated, integrated approach with a defined end-state. We also recommended that the National Coordinator for Security, Infrastructure Protection, and Counterterrorism actively review and guide the growing number of consequence management training and equipment programs and response elements to ensure that individual agencies' efforts (1) leverage existing state and local emergency management systems and (2) are coordinated, unduplicated, and focused toward achieving a clearly defined end-state. More recently, we have noted that rapid program growth, particularly in domestic preparedness programs and public health initiatives, has occurred in the absence of soundly established requirements based on assessments of the threat and risk of terrorist attack involving CBRN. A critical piece of the equation in decisions about establishing and expanding programs to combat terrorism is an analytically sound threat and risk assessment using valid inputs from the intelligence community and other disciplines. Threat and risk assessments could help the government make decisions about how to target investments in combating terrorism and set priorities on the basis of risk; identify unnecessary program duplication, overlap, and gaps; and correctly size individual agencies' levels of effort. Without adequate assessment based on sound input, it would be difficult, if not impossible, to have confidence that the government has properly shaped programs and focused resources to combat and prepare for this complex, emerging threat.

CONCLUSIONS

The executive branch has taken a number of steps toward improving the overall management and coordination of the complex, growing array of agencies' and offices' efforts to combat terrorism. Nevertheless, we see opportunities to improve the overall focus of the nation's efforts to combat and prepare for terrorist incidents. There is a need to reconcile conflicting statements about the CBRN terrorism threat and the lack of connectivity between intelligence judgments and program initiatives. There is also a need for a governmentwide strategy with a defined end-state and priorities, soundly defined requirements based on valid assessments of the threat and risk of terrorist attack, and a comprehensive inventory of existing capabilities and assets. In the absence of these fundamental program elements, there has been significant growth in federally funded consequence management training and equipment programs for first responders and in federal teams, assets, and capabilities to deal with possible CBRN terrorist incidents. Without these program elements, there is little assurance that the nation is investing in the right programs and in the right amounts.

Major contributors to this testimony are Stephen L. Caldwell, Davi M. D'Agostino, and Robert L Pelletier.

Madam Chairman, that concludes my prepared statement. I would be happy to answer any questions at this time.

APPENDIX I APPENDIX I

INFORMATION ON SELECTED FEDERAL TRAINING AND EQUIPMENT PROGRAMS FOR FIRST RESPONDERS

The following summarizes some aspects of selected federal consequence management training and equipment programs designed for state and local first responders to deal with chemical, biological, radiological, and nuclear (CBRN) terrorist incidents.

- DOD: In the Defense Against Weapons of Mass Destruction Act, (Title XIV, P.L. 104-201, Sept. 23, 1996)—commonly known as the Nunn-Lugar-Domenici Act—the Congress authorized DOD to develop and conduct first responder training focusing on terrorist incidents involving CBRN weapons. In designing the Nunn-Lugar-Domenici Domestic Preparedness Program, DOD targeted the 120 most populated U.S. cities to receive this training. Courses are to be delivered to experienced city trainers so they can train rank-and-file first responders. The five-year loan agreement governing the provision of CBRN items and equipment associated with the program requires the cities to repair, maintain, and replace the equipment. DOD plans to transfer responsibility for its domestic preparedness training and equipment program to the Department of Justice by the end of fiscal year 2000.

- Department of Justice: Through the Antiterrorism and Effective Death Penalty Act of 1996, the Congress authorized a second terrorism-related consequence management training program for firefighters and emergency medical personnel. This program, developed in conjunction with the Federal Emergency Management Agency (FEMA), is administered by the Office of Justice Programs. The target audience for this program overlaps with but is not identical to the target audience for DOD's Domestic Preparedness Program. In fiscal years 1998 and 1999, the Congress appropriated $103.5 million to make chemical/biological equipment permanently available to first responders through the Office of Justice Programs. The Department of Justice also is establishing a Center for Domestic Preparedness at Fort McClellan, Alabama. Other Justice-funded centers and training venues related to combating terrorism are at universities, such as Texas A&M and Louisiana State University, and at Department of Energy's (DOE) Nevada Test Site.

- FEMA: Through its National Fire Academy and Emergency Management Institute, FEMA offers training and issues basic course materials. FEMA and its National Fire Academy have long-standing resident and nonresident training programs in emergency management and hazardous materials. FEMA requested about $31 million for fiscal year 2000--a $13-million increase over fiscal year 1999 funding. Of the $31 million, $29 million is to provide grants and assistance related to training, planning, and exercises for state and local responders.

- Environmental Protection Agency (EPA): EPA's Environmental Response Team provides training to federal, state, and local hazardous materials technicians that addresses radiological, biological, and chemical hazards. EPA is adding training to its course that deals with CBRN weapons.

- DOE: DOE sponsors training in how to respond to incidents involving the release of nuclear or radiological substances. The training is made available primarily to communities in which nuclear facilities are located.

- HHS and Department of Veterans Affairs (VA): The Defense Against Weapons of Mass Destruction Act authorized funds for DOD to assist the Secretary of HHS in establishing Metropolitan Medical Response Systems to help improve local jurisdictions' medical response capabilities for a CBRN incident. HHS' Office of Emergency Preparedness has been establishing Systems with trained and equipped local emergency teams in 27cities that also participate in the Nunn-Lugar-Domenici domestic preparedness training and equipment program. VA is involved in training through a contract from HHS. Specifically, HHS is contracting with VA to train 1,100 non-federal National Disaster Medical System hospital staffs to deal with CBRN situations, according to VA officials.

RELATED GAO PRODUCTS

Combating Terrorism: Use of National Guard Response Teams Is Unclear (GAO/NSIAD-99-110, May 21, 1999).

Combating Terrorism: Issues to Be Resolved to Improve Counterterrorist Operations (GAO/NSIAD-99-135, May 13, 1999).

Weapons of Mass Destruction: DOD Efforts to Reduce Russian Arsenals May Cost More and Accomplish Less Than Expected (GAO/NSIAD-99-76, Apr. 13, 1999).

Combating Terrorism: Observations on Biological Terrorism and Public Health Initiatives (GAO/T-NSIAD-99-112, Mar. 16, 1999).

Combating Terrorism: Observations on Federal Spending to Combat Terrorism

(GAO/T-NSIAD/GGD-99-107, Mar. 11, 1999).

Combating Terrorism: FBI's Use of Federal Funds for Counterterrorism-Related Activities (FYs 1995-98) (GAO/GGD-99-7, Nov. 20, 1998).

Combating Terrorism: Opportunities to Improve Domestic Preparedness Program Focus and Efficiency (GAO/NSIAD-99-3, Nov. 12, 1998).

Combating Terrorism: Observations on the Nunn-Lugar-Domenici Domestic Preparedness Program (GAO/T-NSIAD-99-16, Oct. 2, 1998).

Combating Terrorism: Observations on Crosscutting Issues (GAO/T-NSIAD-98-164, Apr. 23, 1998).

Combating Terrorism: Threat and Risk Assessments Can Help Prioritize and Target Program Investments (GAO/NSIAD-98-74, Apr. 9, 1998).

Combating Terrorism: Spending on Governmentwide Programs Requires Better Management and Coordination (GAO/NSIAD-98-39, Dec. 1, 1997).

Combating Terrorism: Efforts to Protect U.S. Forces in Turkey and the Middle East (GAO/T-NSIAD-98-44, Oct. 28, 1997).

Combating Terrorism: Federal Agencies' Efforts to Implement National Policy and Strategy (GAO/NSIAD-97-254, Sept. 26, 1997).

Combating Terrorism: Status of DOD Efforts to Protect Its Forces Overseas (GAO/NSIAD-97-207, July 21, 1997).

STATEMENT OF

CATHERINE H. LIGHT

DIRECTOR

OFFICE OF NATIONAL SECURITY AFFAIRS

FEDERAL EMERGENCY MANAGEMENT AGENCY

BEFORE THE

SUBCOMMITEE ON OVERSIGHT, INVESTIGATIONS, AND

EMERGENCY MANAGEMENT

COMMITTEE ON TRANSPORTATION AND INFRASTRUCTURE

U.S. HOUSE OF REPRESENTATIVES

JUNE 9, 1999

Introduction

I am Catherine H. Light, the Director of the Office of National Security Affairs in the Federal Emergency Management Agency (FEMA). On behalf of FEMA Director James Lee Witt, I thank you for the opportunity to discuss FEMA's terrorism-related activities with the Subcommittee. I will provide an overview of FEMA's responsibilities in terrorism preparedness and response; briefly describe the FEMA role in consequence management; and discuss FEMA programs and activities to better prepare State and local officials for dealing with the effects of terrorist incidents.

FEMA Responsibilities

The basis for FEMA's domestic preparedness and response activities derives from two Presidential Decision Directives. Under PDD-39 and reinforced under PDD-62, FEMA is responsible for the consequence management of domestic terrorist incidents involving weapons of mass destruction (WMD). Under PDD-39, FEMA was given the responsibility to ensure that the Federal Response Plan (FRP) is adequate for consequence management activities in response to domestic terrorist attacks involving WMD. Also, FEMA was tasked to ensure that State response plans and capabilities are adequate and tested.

Crisis Management and Consequence Management

Crisis management focuses on causes and involves activities to address the threat or occurrence of a terrorist incident. It is predominantly a law enforcement function that includes measures to anticipate, prevent and/or resolve a threat or act of terrorism. The lead agency for crisis management is the FBI. Consequence management addresses the effects of an incident on lives and property. It includes measures to protect public health and safety, restore essential government services, and provide emergency relief to governments, businesses, and individuals affected by a terrorist incident. FEMA is the lead agency for consequence management.

Federal Response Plan

In an actual or potential terrorist incident, FEMA uses the structures and resources of the Federal Response Plan (FRP) to manage the Federal consequence management response. The FRP, first published in 1992 and recently updated, has been used in the past several years to respond to numerous disasters and emergencies declared by the President, including the Oklahoma City bombing in 1995, as well as hurricanes, tornadoes, floods and earthquakes. The Plan brings together twenty-six Federal departments and agencies and the American Red Cross to organize Federal disaster response and recovery efforts and coordinate them with an affected State. Most importantly, it provides a known and flexible framework under which local, State and Federal officials can orchestrate their response to a disaster or emergency and make the most effective use of all available resources. FEMA has developed a special annex to the FRP to address the unique requirements involved in responding to a terrorist incident.

Domestic Preparedness Activities

Although the general threat of a terrorist attack exists, FEMA is not in a position to identify the likelihood of a particular incident or the geographic location where an attack may take place. To ensure the broadest coverage in implementing domestic terrorism preparedness activities, we are emphasizing the following key considerations to ensure that:

- State and local first responders and emergency management personnel are the focus of Federal programs.
- Needs of the balance of the nation, particularly local jurisdictions beyond the largest cities and metro areas, are addressed with plans, training, exercises and equipment.
- Initial training is reinforced and sustained with refresher information and updated instruction.
- Existing plans, capabilities and systems are utilized as the foundation for addressing the unique requirements of WMD.

National Domestic Preparedness Office

FEMA Director Witt has been working closely with the Attorney General to better coordinate the interagency efforts for domestic preparedness, including support for the establishment of the National Domestic Preparedness Office (NDPO) to coordinate overall Federal activities with the States and their first responder and emergency management counterparts. In conjunction with the NDPO, FEMA will continue to support preparedness activities in the areas of planning, training, exercises and equipment to further assist the State and local response communities. We also will continue to maintain our traditional links to the States for implementing FEMA programs and activities as we work closely with the FBI to help ensure NDPO activities are fully coordinated with FEMA Regional Offices and the States.

FY 1999 Terrorism Preparedness Budget

In Fiscal Year 1999, FEMA is making some $12.2 million in grants available to State and local jurisdictions for terrorism-related preparedness activities. This includes $8.2 million for State emergency management agencies to support terrorism consequence management planning, training and exercise activities. It also includes $4.0 million for State fire training centers to support delivery of the FEMA National Fire Academy *Emergency Response to Terrorism* training curriculum.

Planning

The responsibility for developing plans and implementing response falls heavily on the States and the local governments. FEMA is applying its experience gained in responding to natural disasters to guide the development of terrorism consequence management preparedness plans and procedures at the local, State and Federal levels. In 1997, we published the FRP Terrorism Incident Annex describing policies and structures of the Federal government for coordinating crisis management and consequence management activities. A fundamental goal of the planning effort is to assist in the development of State and local plans for dealing with WMD contingencies. FEMA grant assistance is being used to enhance planning resources and capabilities at the State and local levels of government. The planning effort is being coordinated with the FBI utilizing existing plans and associated planning structures whenever possible to help ensure that crisis and consequence management plans are in place across the nation.

Training

FEMA has developed and delivered a number of terrorism-related courses for State and local emergency management personnel and first responders. FEMA is utilizing existing programs, networks and facilities to help support the training delivery. In particular, we are using the National Emergency Training Center, which includes the National Fire Academy and the Emergency Management Institute, as well as State fire and emergency management training systems to deliver terrorism-related training to State and local responders. FEMA continues to emphasize the "train-the-trainer" approach to leverage existing capabilities with performance objectives to accomplish training goals.

The National Fire Academy (NFA) has developed courses for first responders in the fire community and other areas. Working with the Department of Justice, NFA developed and has fielded several courses in the *Emergency Response to Terrorism (ERT)* curriculum. The first offering, a Self-Study course, provides general awareness information for responding to terrorist incidents and has been distributed to some 35,000 fire/ rescue departments; 16,000 law enforcement agencies; and over 3,000 local and State emergency managers in the United States.

The Self-Study course is available on the FEMA internet site and will soon be available through the FBI Law Enforcement Online (LEO) intranet site. Other courses in the curriculum deal with Basic Concepts; Incident Management; and Tactical Considerations for Emergency Medical Services (EMS), Company Officers, and HAZMAT Response. Approximately one thousand ERT instructors representing every State and major metropolitan area in the nation will have been trained under this program by July of this year. The NFA is utilizing the Training Resources and Data Exchange (TRADE) program to reach all 50 States and all major metropolitan fire and rescue departments with training materials and course offerings. To date, over 46,000 fire and emergency response personnel have received ERT training.

In addition, FEMA is using the Emergency Management Institute (EMI) to develop and deliver courses such as the Integrated Emergency Management Course on Terrorism that uses a terrorist attack scenario. This 4½-day team-building program includes classroom instruction and a tabletop exercise specifically tailored for emergency response leaders from a particular community or jurisdiction. EMI also has developed and delivered a number of other terrorism-related courses such as incident command, mass fatalities, and emergency response to criminal and terrorist incidents. EMI is utilizing the State Emergency Management Training Officers (STOs) network to distribute materials and instruction to the emergency management community in all 50 States and six territories.

FEMA's course development efforts have enhanced initiatives of both the Department of Justice and the Department of Defense. DOJ has been using the *Emergency Response to Terrorism* curriculum in its training program; the Department of Defense has used the Senior Officials Workshop developed by the EMI in its Domestic Preparedness Program. The benefit added by FEMA funding assistance is that it goes to the States, not directly to local jurisdictions. The States are then able to engage jurisdictions not reached by the DOD and DOJ efforts in terrorism preparedness.

Exercises

FEMA is working closely with the NDPO, the FBI and the States to ensure the development of a comprehensive exercise program that meets the needs of the first responder communities and other response elements. The FBI, in coordination with FEMA, DOD, DOE, HHS, and EPA, and other departments and agencies, will ensure the implementation of a comprehensive terrorism exercise program. State and local involvement in the planning, scheduling, and conduct of coordinated exercises will be a key component of the exercise effort.

Equipment

The NDPO has assumed the lead for developing a list of standardized equipment for the first responder community. FEMA, along with other Federal agencies and first responder communities, is helping to develop the required Standard Equipment List (SEL) to support acquisition of comparable equipment by DOJ for the first responder community. This equipment will conform to appropriate and applicable laws, regulations, and standards, such as those issued by the National Institute for Occupational Safety and Health and the National Fire Protection Association.

FEMA has developed and implemented the Rapid Response Information System (RRIS), which contains an inventory of Federal assets that could be made available to assist State and local response efforts, and a database on chemical and biological agents and protective measures. The inventory is being made available to Federal, State and local officials to assist them in assessing and obtaining the necessary equipment and resources for responding to terrorist incidents involving the use of nuclear, chemical or biological weapons.

Conclusion

Terrorism preparedness requires planning, training and exercising on a regular basis, and appropriate equipment, to ensure maximum readiness to respond to an actual incident. In responding to a terrorism incident, local responders will be the first to arrive at an incident site and may be forced to manage operations at the scene on their own for hours.

From our consequence management perspective, we recognize the extreme importance of viable State and local response plans and capabilities as critical to this effort.

FEMA is committed to work with the Federal interagency community in coordinating our activities and programs as part of the overall Federal efforts. And we are committed to doing everything we can to better prepare the States and local jurisdictions for dealing with this immense challenge.

House Transportation and Infrastructure Committee
Oversight, Investigations, and Emergency Management Subcommittee
Follow-Up Questions from June 9, 1999, Hearing

1. *FEMA's FY 2000 budget request includes $750,000 in order to revise weapons of mass destruction (WMD) responder-training courses. What plans does FEMA have to coordinate these revisions with other federal agencies that offer similar training? Assuming you receive the funding, when would the course revisions be completed and ready for use in responder training?*

These increased resources for FEMA's Emergency Management Institute (EMI) will primarily support instructional delivery of existing FEMA terrorism consequence management training courses, including the *Integrated Emergency Management Course (IEMC) on Consequences of Terrorism* and instructor training for FEMA's terrorism-related courses. With respect to revisions, EMI staff will coordinate closely with other departments and agencies, as they have done in the past, to ensure that doctrine and other course content is accurate and consistent. The EMI deputy superintendent is a member of the National Domestic Preparedness Office (NDPO) Training Committee, and we will coordinate all EMI terrorism training activities (development, revision, and delivery) through the NDPO.

For its part, FEMA's National Fire Academy (NFA) estimates that development of a new course takes 18 to 24 months, depending on course length. A similar timeline is envisaged for revisions to NFA's *Chemistry of Hazardous Materials* course, which will be modified to address chemical and biological agents used as weapons of mass destruction (WMD). Development team members for the revision would include representatives from the U.S. Army's Soldiers and Biological-Chemical Command (SBCCOM) and the U.S. Army Medical Research Institute for Infectious Diseases (USAMRIID).

NFA is also represented on NDPO's Training Committee and coordinates its training efforts with the NDPO. NFA is working with NDPO to explore opportunities for training on the Incident Command System (ICS) and its application to the crisis and consequence management response to terrorism. NFA also is working with the Department of Health and Human Services (HHS) to integrate NFA's *Emergency Response to Terrorism: Tactical Considerations for Emergency Medical Services* course into HHS's curriculum delivery system.

2. *FEMA and the Department of Justice are conducting training in 120 jurisdictions. When will this training be completed? Has FEMA considered moving any training sites because of the proximity to a Department of Defense training site?*

Section 819 of Public Law 104-132, the Antiterrorism and Effective Death Penalty Act of 1996, authorizes the Attorney General, "in consultation with the Director of the Federal Emergency Management Agency, to make grants to provide specialized training and equipment to enhance the capability of metropolitan fire and emergency service departments to respond to terrorist attacks."

The Department of Justice (DOJ) selected the 120 jurisdictions that would be part of its training program. In a 1997 statement of understanding between FEMA and the DOJ, FEMA agreed that it would develop courses and facilitate instructor training for the program through FEMA's National Fire Academy in Emmitsburg, Maryland. NFA has developed a total of six terrorism courses in cooperation with DOJ:

Emergency Response to Terrorism (ERT): Self-Study
ERT: Basic Concepts
ERT: Tactical Considerations - Emergency Medical Services
ERT: Tactical Considerations - Hazardous Materials
ERT: Tactical Considerations - Company Officer
ERT: Incident Management

3. *You are administering grants to address training in the "balance of the nation." What is the legislative authority for these grants? What is the total amount of funding available for these grants? How many such grants has FEMA awarded [in FY 1999]? How many such grants does FEMA expect to award [in FY 2000]? How does FEMA decide who to award these grants to and the amount to award? When do you expect this grant program will be completed?*

FEMA's terrorism-related grants to State emergency management agencies are for consequence management planning, exercises, and training, with an emphasis on planning. Grants to State fire training agencies are for training only.

Apart from appropriations committee language, legislative authority for the grants to State emergency management agencies is, *inter alia*, at Stafford Act paragraph 611(j)(1): "The Director may make financial contributions, on the basis of programs or projects approved by the Director, to the States for emergency preparedness purposes, including the procurement, construction, leasing, or renovating of materials and facilities." The word "including" is construed as "including but not limited to." Authority to make grants to State fire training agencies is in the Fire Prevention and Control Act of 1974, "The Administrator is authorized to provide assistance to State and local fire service training programs through grants, contracts, or otherwise."

For fiscal year 1999, FEMA is providing $8.3 million in assistance to emergency management agencies in the 50 States, the District of Colombia, and Puerto Rico. Funding to a State consists of a base amount (equal for all recipients) and an amount pro-rated among the 50 States based on the number of large local jurisdictions. FEMA is also providing $4 million to the fire training agencies of the 50 States, divided equally among the 50 States.

FEMA anticipates that testing and revising plans for terrorism consequence management response will be a continuing requirement, as will training in this area.

4. *What is the status and schedule of any of the following activities that the Federal Emergency Management Agency may be performing:*

- *an inventory and assessment of existing response capabilities and assets at the federal level;*
- *an identification of response needs at the federal level;*
- *an inventory and assessment of existing response capabilities and assets at the state and local level;*
- *an identification of response needs at the state and local level; and*
- *a written threat and risk assessment of cities or regions of the United States regarding a terrorist attack?*

Since March 1998, FEMA, along with our federal interagency partners, has been engaged in developing a means of providing a rapid response to the consequences of a domestic terrorist incident involving weapons of mass destruction (WMD). One of our efforts has been focused on planning the deployment of federal response resources (teams and assets) in a rapid, logical, and time-phased sequence to assist state and local governments in coping with the devastating consequences of a WMD incident. Federal resources such as agent identification capabilities, medical assistance, decontamination, and pharmaceuticals may be deployed to supplement state and local capabilities. FEMA has developed a database containing a comprehensive list of appropriate federal resources and is in the process of identifying "packages" of the specific federal assets that might be required to respond to a chemical, biological, nuclear detonation, radiological dispersion, or a conventional high explosive (Oklahoma City) event. The resources in these "Force Packages" will be time-phased into the event area in a logical manner. FEMA intends to have five Force Packages–one for each of the above scenarios– developed by the end of this fiscal year. These Force Packages will be included in the database by the end of the calendar year.

Also, the Rapid Response Information System (RRIS) has been operational since December 1997. The system contains several databases on federal capabilities, munitions, chemical, biological and radiological characteristic data, and Federal WMD/terrorism-related training courses. The system also has numerous links to various other internet WMD/terrorism sites and Federal Agencies. The federal capabilities database has information on organizations, their mission, general equipment capabilities, location, and point of contact for further information. A partial version of the system is available to the public on the internet; the complete RRIS is available via a secure, restricted access site.

Regarding State capabilities and needs, the Capability Assessment for Readiness (CAR) establishes a national set of performance criteria which has been accepted by all States as an emergency management baseline. The CAR is a self-assessment process that was completed by all States and Territories in 1997. It involves an assessment of State capabilities in 13 functional areas. However, for the purposes of response capabilities and assets, six functions are crucial: 1.

Resource Management, 2. Planning, 3. Direction, Control & Coordination, 4. Communications and Warning, 5. Operations and Procedures, and 6. Logistics and Facilities. An analysis of the CAR data provides an identification of response needs. For example, on questions involving WMD the analysis indicated that this is an area that requires greater emphasis on planning and equipment. The CAR is currently under revision and will include an enhancement of WMD questions. The next State CAR assessment will be conducted in FY 2000. A CAR assessment for local jurisdictions is currently under development and could be implemented in FY 2000 or FY 2001.

The National Fire Academy also is currently soliciting information from each State's hazardous materials point of contact to identify the number of level "A" hazmat teams (both public and private sector), the number of responders trained to technician level, and related information pertinent to each State.

In regard to assessing the threat of terrorism in the United States, that is a function of the Federal Bureau of Investigation.

5. ***Please provide copies of the last two Reports to Congress submitted pursuant to section 624 of the Robert T. Stafford Disaster Relief and Emergency Assistance Act.***

FEMA thanks the Subcommittee for bringing this reporting requirement to our attention. Should the annual budget submission not suffice, we will work with the Subcommittee to provide the necessary information.

6. *What annual goals and measures developed by FEMA pursuant to the Government Performance and Results Act relate to preparedness against terrorist attack?*

Terrorism is one hazard among many; FEMA's strategic plan includes goals and objectives for improving Federal, State, and local capabilities to respond to all hazards. Thus, there is no terrorism-specific performance measure in the FY 1999 annual performance plan. One performance indicator, the Capability Assessment for Readiness (CAR), includes terrorism-related information. Similarly, follow-on surveys of students receiving FEMA training or evaluation of exercises would include information on terrorism-related training and terrorism-related exercises.

U.S. Department of Justice

Federal Bureau of Investigation

Washington, D.C. 20535

STATEMENT BY

MRS. BARBARA Y. MARTINEZ

DEPUTY DIRECTOR
NATIONAL DOMESTIC PREPAREDNESS OFFICE

BEFORE THE
UNITED STATES HOUSE OF REPRESENTATIVES
TRANSPORTATION AND INFRASTRUCTURE
COMMITTEE

SUBCOMMITTEE ON OVERSIGHT,
INVESTIGATIONS, AND EMERGENCY MANAGEMENT

"PREPAREDNESS FOR TERRORISM RESPONSE"

JUNE 9, 1999

Good morning, Madam Chairman and thank you for this opportunity to speak before distinguished members of Congress and my colleagues regarding the proposed role of the National Domestic Preparedness Office in combating terrorism within the United States.

My intent is to highlight the importance of achieving coordination across the federal government of the various individual agency efforts that provide valuable assistance to states and local communities in preparing them to face the challenge that terrorism presents. As over 40 federal agencies would have a role in the response to a terrorist attack involving weapons of mass destruction, so too are many of these agencies in a logical position to provide various forms of expert assistance to their state and local counterparts -- the men and women of this country whose job it is to save lives and protect the security of our communities if such an event occurs. The mission of the proposed National Domestic Preparedness Office, consistent with the recommendation to the Attorney General by State and local authorities, will be to serve as the central coordinating body for federal programs that can help emergency responders prepare for terrorist incidents, particularly those involving weapons of mass destruction.

Potential Threat of a Terrorist Attack involving Weapons of Mass Destruction

Terrorist events such as the World Trade Center bombing, the bombing of the Alfred P. Murrah Federal building in Oklahoma City, and the pipe bomb at the Olympic Games in Atlanta revealed the united states' increased susceptibility to terrorist assaults. These attacks, coupled with the March 1995 Tokyo subway attack, where the weapon was the chemical nerve agent sarin, exposed the threat of use of WMD within the united states. The threat of WMD use in the United States is real, however, we must not inflate nor understate the actual threat. The United States is experiencing an increased number of hoaxes involving the use of chemical or biological agents perpetrated by individuals wishing to instill fear and disrupt communities. Yesterday's bomb threat has been replaced with a more exotic biological or chemical threat. While the FBI continues to investigate these hoaxes, other on-going investigations reveal that domestic extremists, as well as international terrorists with open anti-U.S. sentiments, are becoming more interested in the potential use of chemical and biological agents.

Examining the increased number of WMD criminal cases the FBI has opened over the past several years highlights the potential threat of use we face. WMD criminal cases are those cases primarily dealing with the use, threatened use, or procurement of chemical and biological materials with intent to harm within the United States. These criminal cases have shown a steady increase since 1995, rising from 37 in 1996 to 74 in 1997, 181 in 1998, and 114 to date for 1999, with three-quarters of these cases threatening a biological release. The biological agent most often cited in 1998 and 1999 was anthrax. Despite the increase in fabricated threats, the WMD threat remains. Since the early 1990s, the FBI has investigated a number of domestic extremist groups and associated individuals interested in procuring or ready to employ chemical or biological agents against innocent civilians. In February 1999, members of a right-wing splinter group were sentenced to 292 months (over 24 years) in prison for threatening to use a weapon of mass destruction against federal officials. These individuals intended to modify a cigarette lighter in order to shoot cactus quills tainted with HIV-blood or rabies.

60-220 00 - 5

It is impossible to eliminate all vulnerabilities in an open society without taking draconian measures that impinge on civil liberties. However, it is possible to reduce susceptibility to WMD terrorist attacks by taking security precautions, remaining vigilant in pursuing WMD terrorist activity, and improving preventive measures, as well as civil preparedness. The FBI is currently undertaking all of these steps. The United States is preparing itself for unconventional threats like WMD terrorism by coordinating federal, state, and local law enforcement and emergency responders in their ability to ferret the fabricated threats and meet the challenges posed by a potential chemical or biological terrorist attack.

As you know, in the past few years, the President of the United States and Congress have taken significant steps to increase our national security and to promote interagency cooperation. Most recently, cooperative efforts against terrorism have been extended to include state and local agencies and professional and private sector associations as well.

For example, in the preparation of the Administration's Five-Year Interagency Counterterrorism and Technology Crime Plan, the Attorney General directed the Department of Justice, Office of Justice Programs, to host a meeting of individuals who represent the various emergency response disciplines that would most likely be involved in the response to a terrorist event. More than 200 stakeholders representing local and State disciplines of fire services and HAZMAT personnel; law enforcement and public safety personnel; emergency medical and public health professionals; emergency management and government officials; and various professional associations and organizations attended the two-day session.

Collectively, they made recommendations to the Attorney General; James Lee Witt, Director of FEMA; Dr. Hamre, the Deputy Secretary of Defense, and other Federal officials on ways to improve assistance for state and local communities. These recommendations have been incorporated in the Administration's Five-Year Plan mentioned above.

The most critical issue identified by stakeholders was the need for a central federal point of coordination. Due to the size and complexity of both the problem of terrorism and of the federal government itself, it was no surprise that the many different avenues through which aid may be acquired, by state and local officials, and the potential inconsistency of those programs was deemed to be simply overwhelming. In essence, the federal government, though well intentioned, was not operating in an optimal manner nor was it effectively serving its constituents with regard to domestic preparedness programs and issues in an optimal manner.

State and local emergency response officials made a strong recommendation to the Attorney General for the coordination and integration of all federal assistance programs that reach state and local agencies for terrorism preparedness. In heeding that recommendation and seeking to increase the efficiency and effectiveness of federal support programs that provide grants for equipment, training, exercises, and information sharing, the Attorney General proposed the establishment of the National Domestic Preparedness Office.

In proposing the establishment of the NDPO, the Attorney General consulted the National Security Council, Department of Defense, Federal Emergency Management Agency, Department of Health and Human Services, and other relevant agencies regarding the creation of a single coordination point within the federal government to better meet the needs of the Nation.

Mission of the NDPO

The NDPO, if approved, will provide a forum for the coordination of all federal programs that offer WMD terrorism preparedness assistance for State and local officials. Through such coordination, it is believed that the vital efforts of the Office of Justice Programs' Office for State and Local Domestic Preparedness Support, the Federal Emergency Management Agency (FEMA), the Environmental Protection Agency (EPA), the Department of Health and Human Services (DHHS), the Department of Defense (DoD), the National Guard Bureau (NGB), the Federal Bureau of Investigation (FBI), and other agencies will better serve the States and local communities of this country.

It is intended that the NDPO will serve as a much needed clearinghouse to provide information to local and state officials who must determine the preparedness strategy for their community. In keeping with Stakeholders' requests, the NDPO will also provide a forum for the establishment of agreed-upon standards to guide the execution of federal programs.

Federal participants that will serve in a full-time capacity at the NDPO, once approved, will include the Federal Emergency Management Agency, the Department of Defense, the National Guard Bureau, the Department of Energy, the Department of Health and Human Services, the Environmental Protection Agency, the Office of Justice Programs, and the Federal Bureau of Investigation. We have also received commitments from other agencies including the U.S. Coast Guard, the Department of Veterans Affairs, and the Nuclear Regulatory Commission, to provide personnel in the future.

Stakeholders also cited the need for formal representation of state and local officials with the federal agencies in the form of an Advisory Board to guide the development and delivery of more effective federal programs. Federal agencies agree that their participation is critical to the whole process of domestic preparedness. Therefore, in addition to the Advisory Board, it is anticipated that when fully staffed, approximately one-third of the NDPO will be comprised of State and Local experts from various disciplines.

Stakeholders identified six broad issue areas in need of coordination and assistance. These areas are: Planning; Training; Exercise; Equipment Research and Development; Information Sharing; and Public Health and Medical Services. I would like to highlight how the proposed NDPO would address each of these areas.

In the area of Planning, the NDPO would facilitate the distribution of the United States Government Interagency Domestic Terrorism Concept of Operations Plan and other Planning guidance for State and Local communities. The benefit of such guidance is to explain to state and local planners the logistics of how federal assets may be included in their local emergency response plans.

In the area of Training, the NDPO would continue the DoD initiative to establish and maintain a compendium of existing federal training courses available to emergency responders.

It would also establish a mechanism to ensure federal training programs comply with national standards such as those issued by the National Fire Protection Association and Occupational Safety and Health Administration. Finally, it would develop a national strategy to make sustained training opportunities and assistance available to all communities and States. For example, the Office of Justice Programs Office for State and Local Domestic Preparedness Support will incorporate into the training programs that it supports standards that have been coordinated through the NDPO process.

In connection with the Information Sharing program area, the NDPO can implement a mechanism to facilitate access by personnel outside law enforcement to information that may be important for preparedness and consequence management. Internet web-sites, both public and secure have been proposed for the sharing of public safety information. Links to several existing web-sites may also be built.

In the Exercise program area, the NDPO will formally adapt a military software application for civilian use to track the lessons learned during exercises and actual events. The NDPO will provide this tool to participating communities and will maintain an After-Action Tracking database for the repository and review of all lessons that might assist other communities.

In the Equipment/Research and Development program area, the NDPO has established a Standardized Equipment List which has been incorporated into the grant application kits used by the Office of Justice Programs. The NDPO would, again, serve as a clearinghouse for product information provided by private vendors and testing data provided by approved testing facilities to promote synergy and avoid costly duplication in the area of federal research and development.

In the Health and Medical program area, the NDPO, under the guidance of the Public Health Service of the Department of Health and Human Services would coordinate efforts to support Metropolitan Medical Response Systems, pharmaceutical stockpiling, the establishment of a nationwide surveillance system to improve the identification of infectious diseases and the integration of the public and mental health care community into WMD response plans.

Thus far, two conferences have been held and have been attended by representatives from Federal, State and local agencies to promote interaction. Each time, the Attorney General was presented with an overview by several communities of their cooperative efforts, which illustrated the growing cooperation between all levels of government to address the preparedness needs of this Nation to deal with a major terrorist event, including those that involve WMD.

I thank you for the opportunity to speak to you today. As the Attorney General has recently said, "The actions of the first people on the scene will make the difference between life and death. The key is to work together in a partnership among federal, state and local communities to prepare a coordinated response that saves lives and provides for the safety for all involved". I stand ready to respond to any questions you may have.

U.S. Department of Justice

Federal Bureau of Investigation

Washington, D.C. 20535-0001

October 12, 1999

Honorable Tillie K. Fowler
Chairwoman, Subcommittee on Oversight,
Investigations, and Emergency Management
Committee on Transportation and Infrastructure
House of Representatives
Washington, D.C. 20515

Dear Madam Chairwoman:

I am writing in response to your letter to Barbara Martinez dated June 21, 1999, concerning the Subcommittee's questions on certain Department of Justice preparedness activities against terrorist attacks.

We apologize for the delay in providing you with responses to these questions; however, as you are aware, the decision to formally authorize the establishment of the National Domestic Preparedness Office (NDPO) is currently pending before Congress. As such, we believe it would be premature to provide information beyond that which was provided to the Subcommittee in June. With this in mind, your questions numbered 2 and 3, which relate to future plans and staffing schedules of the NDPO, cannot be accurately addressed at this time.

If approved, the NDPO will be an interagency effort to enhance coordination among federal programs offering terrorism preparedness assistance to states and local communities. Separately, initiatives are underway among the federal departments and agencies with regard to preparedness at the federal level. The answers to your questions numbered 1 and 4 fall into this category and are provided in the enclosed document.

We appreciate the Subcommittee's interest in the proposed National Domestic Preparedness Office and look forward to working with you in the future.

Sincerely yours,

NG

Neil J. Gallagher
Assistant Director
National Security Division

Enclosure

1. What is the status and schedule of any of the following activities that the Department of Justice may be performing:

a) An inventory and assessment of existing response capabilities and assets at the federal level:

Pursuant to an NSC directive, FEMA is taking the lead to develop a Timed Phase Force Deployment Database which would inventory and identify existing response capabilities and assetsat the federal level. DOJ, through the FBI is participating as an interagency partner. The development of a comprehensive database along with FEMA is being discussed. A completion date is targeted for early FY 2000.

Another ongoing initiative coordinated by the NSC WMD Preparedness SubCommittee requests the interagency to address the following issues:

-Identify the specialized federal teams that could respond to WMD incidents and determine whether more teams are needed.

-How do we determine when a metropolitan area is trained and therefore ready to respond?

-What should be the U.S. policy on assisting and planning with metropolitan areas?

These issues are currently being addressed by the interagency with DOJ as an integral participant.

b) An inventory and assessment of existing response capabilities and assets at the federal level:

The FBI and OJP are coordinating efforts to develop a comprehensive threat/risk/needs assessment methodology to be utilized at the state level. This will be a self assessment tool which will direct states to inventory and assess their own capabilities and assets in order to determine needs. This model will be developed and tested by September 20, 1999. An actual delivery date has not yet been established.

c) An identification of response needs at the federal level:

This is addressed in the initiatives discussed in 1 a).

d) An identification of response needs at the state and local level:

The identification of response needs at the state and local level is being addressed in the initiative detailed in question 1 b).

e) A written threat and risk assessment of cities or regions of the United States regarding a terrorist attack.

A written threat and risk assessment of cities or regions of the United States regarding a terrorist attack is being developed as part of the initiative detailed in question 1b.

2. What is the schedule for the NDPO to be fully staffed and in its permanent offices?

The decision to formally authorize the NDPO is currently pending before Congress. As such, we believe it would be premature to provide information beyond that which has been provided to the Subcommittee in June. With this in mind, the staffing schedule cannot be accurately addressed at this time.

3. What products (e.g. compilation of training programs) does the NDPO plan to produce in the next 18 months and when during this time period do you expect they will be completed?

The decision to formally authorize the NDPO is currently pending before Congress. As such, we believe it would be premature to provide information beyond that which has been provided to the Subcommittee in June. With this in mind, the planning schedule cannot be accurately addressed at this time.

4. What annual goals and measures developed by Department of Justice pursuant to the Government Performance and Results Act relate to preparedness against terrorist attack?

Annual goals and measures which relate to preparedness against terrorist attack are specified in the FBI's National Security Division, Domestic Terrorism Section Program Plan for 1998 - 2003. This classified document may be obtained through appropriate channels if necessary.

Congressman James L. Oberstar
Ranking Democratic Member
Committee on Transportation and Infrastructure

Subcommittee on Oversight, Investigations and Emergency Management
Hearing on Preparedness Against Terrorist Attacks
June 9, 1999

I would like to commend Chairwoman Fowler for calling this hearing on the state of domestic preparedness against terrorist attacks involving nuclear, biological, and chemical weapons. Additionally, the witnesses will address federal programs that train and support State and local officials.

With the recent increase of terrorist acts in the world, and the availability of weapons of mass destruction (WMD) the U.S. must increase its commitment to ensuring that its citizens are protected from a terrorist attack.

In 1996, the Department of Defense was charged, under the Weapons of Mass Destruction Act of 1996, with enhancing the domestic response to terrorism by providing training and technical assistance to State and local officials nationwide.

At about the same time, Congress passed the Anti-terrorism and Effective Death Penalty Act of 1996 charging the U.S. Attorney General, in consultation with the Director of FEMA, to make grants and provide specialized training to State and local officials nationwide.

However, it appears that following enactment of these laws there was inadequate coordination and consultation among the federal agencies. This produced some fragmented training programs which confused State and local officials.

In response to this criticism and in an effort to alleviate confusion while providing assistance to State and local officials by providing a one-stop-shop, the U.S. Attorney General created the National Domestic Preparedness Office (NDPO) within the FBI. The NDPO could serve as the interagency coordinator and be the single point of contact for State and local officials during a terrorist attack.

Yet, some have already criticized the NDPO because it may not have the authority to eliminate duplicative programs or rationalize the various independently run federal programs. This hearing is our opportunity to further explore that issue and hear testimony on the benefit of providing the NDPO with that authority. It is important for the federal government to have some clear goals and strategies on preparing for the possibility of a terrorist attack. If we do not have a clear plan we may waste valuable resources and lack the preparation necessary to successfully combat a terrorist attack.

I welcome the witnesses and look forward to their testimony.

To strengthen and promote cities as centers of opportunity, leadership, and governance.

National League of Cities

1301 Pennsylvania Ave., N.W.
Washington, D.C. 20004-1763
202-626-3000
Fax: 202-626-3043
Internet: www.nlc.org

National League of Cities 1924 75 years 1999 Opportunity, Leadership, Governance

Testimony of Ann Simank

Council Member, Oklahoma City, Oklahoma

On behalf of

The National League of Cities

Before the

House Subcommittee on Oversight, Investigations and Emergency Management

June 9, 1999

Testimony of Ann Simank
Council Member, Oklahoma City, Oklahoma
On behalf of the National League of Cities
Before the
House Subcommittee on Oversight, Investigations
and Emergency Management
June 9, 1999

Opening

Good morning. Before sharing with you the positions of the National League of Cities (NLC) on domestic terrorism and related issues, I would like to thank the Subcommittee for providing this forum for discussion of the importance of domestic preparedness in the event of a terrorist attack. I am Ann Simank, Council Member from Oklahoma City, Oklahoma, and I am testifying for the National League of Cities, I have been a member of NLC's Public Safety and Crime Prevention Policy Committee for three years. This NLC policy committee has given considerable attention recently to the many issues related to domestic terrorism.

I'd like to summarize our experience in Oklahoma City. On April 19, 1995, at 9:02 a.m., Oklahoma City, my hometown, was devastated by a horrendous terrorist attack, the bombing of the Alfred P. Murrah Federal Building. That fateful morning, 168 lives were lost, thousands were seriously injured, 30 children became orphans, and approximately 300 businesses were destroyed leaving hundreds unemployed. Many lives were literally in chaos. This senseless act shattered my city, all in a matter of seconds.

Were we prepared? I don't know if cities can ever be fully prepared, but today I want to share Oklahoma City's experience with you. Our city had been selected for FEMA's Integrated Emergency Management Course in Emmitsburg, Maryland, in July 1994, less than a year before the bombing. All department heads attended as well as two or three key people from their departments. Numerous representatives from private utilities, private businesses and non-profit organizations also attended. Oklahoma City was well represented at this specialized training event, and we believe it was extremely helpful.

However, Oklahoma City faced difficult challenges on the morning of April 19th. This was the largest terrorist attack ever seen in America, and to our knowledge, no one had ever had to perform rescue and recovery while at the same time effectively deal with a crime scene.

Immediately after the bombing, our mayor, police chief, fire chief, and the local special agent in charge of the FBI got together and made decisions as to how Oklahoma City should proceed. The FBI would be in charge of the crime scene, the Oklahoma City police would seal the perimeter and assist the

FBI, and fire would handle rescue and recovery. Our incident command center was in place, local agreements were made, and everyone was clear as to their role. Within those first few hours, lives were saved, evidence was found, and the system was working well.

Approximately 15 hours later, the first FEMA team flew in to offer assistance. FEMA's normal recovery efforts are to assist in natural disasters – tornadoes, floods, hurricanes and earthquakes, rather than working a crime scene. Relationships quickly became strained between agencies that represented disaster recovery versus agencies that represented law enforcement and criminal investigations. Even some of the FEMA trained USAR Teams concerned themselves with turf battles.

Fortunately, Oklahoma City's incident command system had been up and working since that morning making it difficult for "outsiders" to penetrate or change it. Ultimately, these questions of "who is in charge" were worked out, and our city's system prevailed.

What did we learn in Oklahoma City?

- Damage starts at the point of impact and local responders are the first and only responders. This is true for many hours after an incident.

- Any resource made available to cities can be of benefit if complications are ironed out first and their response is timely.

- Training is essential since cities and towns will be alone the first few hours after an attack.

- Turf issues should not happen; a clear line of command and protocol should be established at the federal level (National Strategic Plan).

- Money should be made available for training and equipment at the local level.

- Transportation for specialized teams needs to be included in a National Strategic Plan. Many USAR Teams had to wait long hours before a military transport could get them Oklahoma City. Frankly, our city has no confidence that military transport will be able to bring specialized teams within four hours as indicated by some federal agencies.

Don't forget, it is the local response that will be first on the scene, first to assess, and the first to take action. Please, see that the resources get to where they need to be.

The acts of terrorism in Nairobi and Dar es Salam and the resulting devastation, make clear the ease with which a city can be shattered. We learned from the terrorist attacks on my city and New York City that cities are the most critical element in any national strategy to confront domestic terrorism, for it is at the heart of cities that terrorist will seek to strike. And it is the nation's cities that bear the heaviest burden to prepare and respond. Ensuring that cities and towns have the unified resources of state and the federal government will be critical to how effective we can be.

The National League of Cities represents 49 state leagues, 135,000 local elected officials and 16,000 direct and indirect member cities. NLC was established in 1924 by the state municipal leagues to represent the interests of cities at the federal level.

The challenges we face to prepare state and local governments and their first responders to deal with terrorist incidents are complex and multifaceted. Whether chemical or biological agents are used, whether the next strike is one involving cyber-terrorism or more conventional Weapons of Mass Destruction (WMD), our cities must be prepared for a variety of threats. Large scale evacuations, public health emergencies, information flow and management, establishing viable communication linkages between federal, state and local responders, the detection of explosives, and identifying other hazardous materials are just a few of the challenges cities will have to be prepared to face in the event of credible terrorist threats or actual terrorist events. To do this, state and local emergency managers and first responders must have proper training, equipment and support from the federal government.

For federal support to be effective, a comprehensive national strategy must be developed which builds on existing federal, state and local emergency preparedness and response infrastructure and relationships. At the core of this national strategy must be the objective of establishing a uniform command and control protocol. This protocol should seek the maximum cooperation of crisis consequence management between and among federal agencies, state and local first responders and the medical community. Additionally, a thorough planning process should be initiated and coordinated by the federal government with the participation and input of state and local elected officials and their emergency managers. It is essential that this process clearly identify federal agency roles, responsibilities, and resources, and give emergency managers one point of contact to activate federal support and to facilitate the exchange of information on all federal programs, training and resources available to meet terrorist threats in our cities and towns. More importantly, this one point of contact should be with a well-informed government official and not some recorded message. It should not change every six months.

We are most appreciative of the federal government's attempts to quell the confusion in this area and establish the National Domestic Preparedness Office (NDPO). At a recent conference,[1] the NDPO outlined its methods for addressing the issue of domestic terrorism preparedness, including programs dealing with planning, information and intelligence sharing, training, exercises, equipment and health and medical. The NDPO also seeks to involve local government by creating a state and local government advisory group that reports directly to the NDPO Director. It is critical that local elected officials, as well as emergency managers, are included in this advisory group. It is also important that the local participants represent cities located in different geographical regions and of various sizes.

Although NLC commends the work of the NDPO thus far, the problem with this latest federal effort is twofold. First and foremost, many departments within other government agencies such as DoD, FEMA and the Justice Department have "laid claim" to one or more of the programs I just enumerated for you. This has created a federal power struggle involving billions of dollars and inordinate amounts of wasted time, energy and resources. True preparedness at the state and local level cannot achieve its greatest potential unless and until the squabbling at the federal level stops.

---

[1] TOPOFF Exercise Planning Conference, May 20-21, 1999

Second, I ask you, what good are any of NDPO's efforts if it cannot receive the funding it needs to implement its objectives and it duplicates the efforts of other federal agencies? What message is sent to state and local first responders when the federal government cannot even coordinate its own efforts?

Importance of a National Policy/Goal

The potential for domestic terrorist events is forcing the federal government to prepare the nation as a whole for such attacks. While there has been an increase in media attention, as well as in federal funding for a variety of programs and projects, NLC believes that these efforts, to date, though well intended, are not achieving their potential due to the absence of a clear national strategy. It is imperative that there be clear lines of authority and coordination between federal agencies regarding their programmatic, operational, and policy activities. Currently, more than 43 federal agencies have roles in handling the challenges related to domestic terrorism. With the creation of the Justice Department's Office for State and Local Domestic Preparedness Support and the NDPO, the number continues to increase.

During a hearing last fall before the House Government Reform and Oversight Committee on domestic preparedness, it was evident that the federal government lacks clearly defined goals. Although the NDPO has been designated recently to be the lead federal agency, coordination of the various federal programs and resources currently available to state and local governments, and cooperation between the federal agencies has not yet become a reality.[2] Witnesses at this hearing highlighted a variety of key problems which included: overlapping federal initiatives, inadequate operational training, and failure to provide training opportunities in many areas of the country. NLC is pleased that these agencies have identified the importance of a lead agency. Our members hope that the National Domestic Preparedness Office, established by the FBI, will succeed in organizing federal resources. It is essential that the various federal agencies involved work cooperatively and plan for the most effective use of the limited resources available. We believe that if these efforts are conjoined with a clearly defined national strategy, state and local governments, and their emergency managers will be able to establish corresponding goals and policies to ensure the execution of domestic preparedness.

Importance of Local/State Preparedness

Initial federal efforts to develop plans, programs, and training initiatives have lacked substantive input from state and local levels of government. But it is unquestionable that cities will have the most urgent responsibility for initial responses, as well as long term recovery. We already have experienced this with the bombing of the World Trade Center in New

[2] The National Security, International Affairs and Criminal Justice Subcommittee (Chairman Hastert, R-Ill) of House Government Reform and Oversight Committee. Held October 2, 1998.

York, as well as the bombing in Oklahoma City. In the event of a terrorist attack, the most critical services and individuals are the municipal emergency responders. Emergency medical personnel, along with fire fighters and law enforcement personnel, are the first to arrive on the scene. Federal support will not be available during the first hours of a terrorist attack. It is important that these first responders, when possible, are trained together on the ground in their local jurisdictions and regions using their own equipment and working with and augmenting the existing emergency response networks. Joint exercises combining local fire fighters, law enforcement personnel, and medical teams will increase preparedness and aid regional cooperation. In this way, they will be able to make practical decisions while training and anticipate what actions must be taken in the aftermath of a terrorist event.

It is critical that local first responders have the equipment and resources they will need to respond. Specialized equipment should be on hand for training exercises and to establish an immediate resource in the event of an attack. State and local governments and their emergency management teams need to know where to obtain such equipment and how to actually acquire it. NDPO assures us that it will establish a program for procurement of and rapid access to vital equipment, as well as for keeping up with the latest technology. Currently, this information is fragmented and difficult to obtain.

The most important factor underlying these various challenges is the importance of communication before, during and after events. Terrorist incidents will require a number of local, state, and federal agencies to cooperate. The lead federal agency should serve as both a coordinator of programs as well as an information clearinghouse to facilitate domestic preparedness across the nation at all levels of government. Timely information and guidance from the lead federal agency will be imperative if cities and towns are to respond immediately and appropriately to a local terrorist incident.

NLC's key recommendations for domestic preparedness include:

1) Development of a national domestic terrorism strategy, clear policies and comprehensive plans to coordinate the roles, responsibilities and resources of federal agencies in support of effective state and local responses to terrorist threats and events.

2) Designation of a lead federal agency that actually serves as the central coordinator and information clearinghouse on all available federal programs and resources.

3) Training, equipment, and resources provided by the federal government; especially regionally based training that builds on shared emergency response networks.

4) A policy for sharing certain classified information on threats or potential threats of terrorism with local law enforcement agencies on a need-to-know basis.

5) The federal government should include local governments in the federal planning process and operations relative to issues in their jurisdictions, and target scarce federal resources in localities that have high profile public or private targets.

Closing

The National League of Cites hopes that our recommendations will aid in establishing a clear delineation of authority and help create clear lines of communication between federal, state, and local governments. We have experienced domestic terrorism in cities; unfortunately, we almost surely will again. If domestic terrorist attacks do occur, cities are committed to respond immediately and effectively and must work with the federal government and our respective state governments if we are to protect our communities.

Again, the National League of Cities would like to thank the Subcommittee for holding this hearing and we look forward to working with all stakeholders to address the issues discussed during this conference.

Statement of the Honorable Lee Terry

Thank you Chairwoman Fowler,

We are here today to hear from representatives of various government agencies about terrorism, the potential for future terrorist attacks, and how the United States is prepared—or preparing—to deal with such attacks.

In recent years, citizens of both New York City and Oklahoma City have had to encounter brutal examples of terrorism and the problems that result from such an attack. Bombings, hijackings or gas attacks are dreadful events. Confusion on how to react to such an event can add to the devastation. Mechanisms must be in place that can be triggered when terrorist attacks take place. The question is: "Who and where should local officials turn for answers and help?"

This year alone, 40 Federal agencies are spending a total of $1 billion to prepare for terrorist attacks. However, there appears to be substantial duplication among these agencies and there is no central point from which to coordinate all of these efforts. For example, if a terrorist attack takes place on a commercial airliner, three different agencies could be involved—one agency if the plane is still parked at the gate with its door closed, another agency if the plane is at the gate with its door open, and a third if the plane is in the air. Since the ultimate goal is to bring an end to this situation, it seems that one agency trained in all facets of response is a better solution rather than three agencies or more.

The fear of future terrorist attacks is REAL. We need to look at our current system and evaluate its effectiveness.

In an era of strained Federal budgets, we should be working to get the biggest "bang for our buck." More money is not always the answer. We must look at what these 40 agencies are doing to prepare for terrorist attacks and see where there may be duplication of effort. We must also consider consolidating these functions under one umbrella, to help better coordinate efforts—both among federal agencies and with local officials who are usually our first line of defense.

I look forward to hearing the testimony of these witnesses.

Congressman James A.Traficant, Jr.
Ranking Democratic Member
Subcommittee on Oversight, Investigations and Emergency Management
Hearing on Preparedness Against Terrorist Attacks
June 9, 1999

I would like to begin by thanking Chairwoman Fowler for calling this very important and timely hearing on the state of domestic preparedness against terrorist attacks involving nuclear, chemical or biological weapons. Today, we shall hear from witnesses who will address federal programs that train and support State and local emergency response officials.

Many of you are familiar with my interest in improving the security of federal buildings. The bombings of the World Trade Center in 1993 and the Murrah Federal Building in 1995 certainly raised serious concerns about domestic vulnerability to terrorist attacks. What many people do not know is that there was only one contract security guard on duty, who was responsible for three federal buildings the day of the bombing in Oklahoma city. My bill, H.R. 809 will make long-overdue changes to the Federal Protective Service, the agency responsible for securing more than 8,300 federal buildings across the country.

To address concerns about the increase in the availability of weapons of mass destruction (WMD), Congress passed the Defense Against Weapons of Mass Destruction Act of 1996 (Nunn-Lugar-Domenici or NLD Act). The bill instructed the Department of Defense (DOD) to enhance domestic preparedness for responding to terrorist use of WMD. The NDL Act mandates that DOD provide training and expert advice to emergency response personnel and lend equipment to State and locals. This program is called the Domestic Preparedness Program (DPP) began in 1998 and is approximately half completed.

Six months after the NLD Act was enacted Congress passed the Anti-Terrorism and Effective Death Penalty Act of 1996. It directed the Attorney General in consultation with the Director of FEMA, to make grants to provide specialized training and equipment to enhance metropolitan fire and emergency service capabilities. FEMA is the lead federal agency in dealing with the consequences of a terrorist attack.

Unfortunately, with what appears to be a limited amount of consultation and coordination between these federal agencies, the

Department of Defense initiated its DPP program by setting-up training and related programs all over the country. Also, during the same time period, FEMA and the several other Federal agencies set-up similar training and WMD programs nationwide.

Initially, some of these federal WMD preparedness programs were fragmented and mostly independent of one another. Some have felt that this lead to the duplication of training, duplication of response teams, and confusion among State and local officials.

Clearly, adding to the confusion was not the goal of any of these programs. As a response to the criticism by State and local officials that the federal training programs and other assistance were not well coordinated, the U.S. Attorney General is establishing within the FBI a new office called the National Domestic Preparedness Office (NDPO).

The NDPO is set-up to be a "one-stop-shop" in an effort to alleviate and reduce confusion while providing assistance to State and local officials. The NDPO will be the single point of contact for State and local officials requesting terrorist related information

.ance while providing a forum for federal interagency
.ination.

Although the NDPO is a valuable asset, the office may not have the adequate authority to rationalize the various independent programs. Additionally, it does not have the authority to eliminate duplicative programs. This hearing may provide us with the opportunity to explore the possibility of providing the NDPO with the authority it needs to truly eliminate confusion while providing quality assistance to State and local officials.

Madame Chairwoman, we must ensure that Federal, State and local officials are well prepared to combat terrorism and that we provide the NDPO with the necessary tools to ensure that we are prepared to effectively respond to terrorist use of weapons of mass destruction. If we do not, we may increase our risk of not being adequately prepared for a terrorist attack and that is not acceptable.

I welcome the witnesses and look forward to hearing their testimony.

O